Errors, Blunders, and Lies:
How to Tell the Difference

ASA-CRC Series on
STATISTICAL REASONING IN SCIENCE AND SOCIETY

SERIES EDITORS

Nicholas Fisher, University of Sydney, Australia

Nicholas Horton, Amherst College, MA, USA

Deborah Nolan, University of California, Berkeley, USA

Regina Nuzzo, Gallaudet University, Washington, DC, USA

David J. Spiegelhalter, University of Cambridge, UK

PUBLISHED TITLE

Errors, Blunders, and Lies: How to Tell the Difference
David S. Salsburg

Errors, Blunders, and Lies:
How to Tell the Difference

By

David S. Salsburg

CRC Press is an imprint of the
Taylor & Francis Group, an **informa** business

CRC Press
Taylor & Francis Group
6000 Broken Sound Parkway NW, Suite 300
Boca Raton, FL 33487-2742

International Standard Book Number-13: 978-1-4987-9578-4 (Paperback)
978-1-138-72698-7 (Hardback)

Library of Congress Cataloging-in-Publication Data
Names: Salsburg, David S., 1931-
Title: Errors, blunders, and lies / David S. Salsburg.
Description: Boca Raton : CRC Press, 2017.
Identifiers: LCCN 2016050456 | ISBN 9781498795784 (pbk. : alk. paper) |
ISBN 9781138726987 (hardback : alk. paper)
Subjects: LCSH: Errors, Scientific. | Science—Statistical methods. | Missing observations (Statistics) | Science—Methodology.
Classification: LCC Q172.5.E77 S38 2017 | DDC 519.509—dc23
LC record available at https://lccn.loc.gov/2016050456

Visit the Taylor & Francis Web site at
http://www.taylorandfrancis.com

and the CRC Press Web site at
http://www.crcpress.com

This book is dedicated to our grandchildren: Matthew and his wife, Amy, Benjamin, Nathan, Joshua, Rebecca, Zachary, Patrick, Ryan, and Joseph. May they have as much fun in their chosen careers as I did in mine.

Contents

Preface

JOHN TUKEY WROTE THAT the fun of being a statistician is that you get to play in other people's backyards. By this, he meant that statistical models have proven so useful that they can be found in all branches of science, in law, in historical research, in Biblical scholarship, and in fields like survey research that would not have existed if there were no statistical models. Statistics bloomed in the twentieth century, playing a role in almost every branch of science. The modern computer has made it possible to use extremely complicated statistical models even though they may require thousands or even millions of mathematical calculations. As a result, statistical methods have a role to play in the twenty-first century, wherever computers come into use.

For more than 50 years I have been involved in statistics. I've had the pleasure of digging around in the mud and muck of many backyards. In this book, I present a few examples of statistics in use, some of them based on my own experiences, some based on experiences of others who have been romping around in different backyards.

I slid over to statistics from graduate work in mathematics. This gave me a solid foundation to understand and use statistical models, a foundation that included courses in measure theory and advanced calculus. However, I realize that few people have the time or inclination to look into Radon-Nikodyn derivatives or Fourier transforms. This book explains the pleasures of statistics to someone who has not pursued math beyond high school algebra. My wife, Fran, who is not a statistician, has been at my side as I wrote this book, pointing out where I have introduced incomprehensible jargon and suggesting more ordinary words to use instead.

One way of looking at statistical models is to note that we live in a world that is not quite "right." Our attempts to measure things are bedeviled with small errors. Our attempts to understand our world are blocked by blunders. And, in some cases, people have been known to lie. As I played in those backyards, I've had to identify the nature of errors, block the effects of blunders, and figure out who are the liars. Come along as I show you what happens in other people's backyards.

Acknowledgments

I WROTE THIS BOOK TO "explain" complicated ideas in what I hope would be an "easily understood" way. However, what I think is "easily understood" may, in fact, be much too convoluted for some readers. Also, in "explaining" complicated ideas based on mathematical formulas, there is always the possibility that the "explanation" would be considered in error by someone conversant in the field.

I am grateful to David Grubbs of Chapman & Hall who found perceptive and critical readers and was the midwife for the final version of this book. I wish to thank the following who read all or portions of this book and provided criticisms that have made the final version much better: Nicholas Fisher, Nicholas Horton, Regina Nuzzo, David Spiegelhalter, Deborah Nolan, Joseph Hilbe, Joseph Blitzstein, and Ann Cannon. And, I thank the Taylor & Francis Group for converting my typescript into this book.

I

The Transit of Venus

H OW FAR IS IT from the Earth to the Sun?
       This has been a question posed by philosophers and
scientists since the time of the ancient Greeks. It plays an important role in today's exploration of the solar system. The distance
from the Earth to the Sun is called an astronomical unit (AU).
The planet Jupiter, for instance, is 5.46 AU from the sun (five and
a half times the distance from the Earth to the Sun). But, how long
is an AU?

In 1716, a British astronomer, Edmund Halley (Figure 1.1),
described a way this could be measured. Astronomers had used
Newton's laws of motion to calculate the orbits of the known
planets and shown that every 113 years, the relative positions of
the planet Venus and the Earth would put Venus between the
Earth and the Sun. With the best of telescopes, astronomers
could determine the time it took for the black dot that was Venus
to cross the face of the Sun. This time would differ depending
upon where you were on the Earth. So, Halley pointed out, if we
could measure that time from two different places on the Earth
that are very far apart, we could use those two different times,
the relative distance of the orbits of Earth and Venus, and a complicated series of calculations to determine the distance from the

FIGURE 1.1 Edmund Halley (1656–1742) who proposed that the distance from the Earth to the Sun could be determined by timing the transit of Venus from two widely separated places on the Earth. (Courtesy of istock.com.)

Earth to the Sun. Reference [1] describes how one could estimate the AU from two such points on Earth, assuming only the knowledge of high school geometry.

When Venus crosses the face of the Sun, it does it twice, 8 years between each transit. The next expected transits of Venus would occur in 1761 and 1769.

Halley died in 1742, but his proposal lived on. As the year 1761 approached, groups of natural philosophers prepared for the event in Russia, Austria, Norway, France, and England. They prepared small lightweight telescopes that could be carried easily. They planned to use the telescopes to cast images onto paper, where they could watch the passage of the black dot of Venus crossing the face of the Sun. (It is impossible to stare directly at the Sun through a telescope without damaging the eye.)

In 1761, most of the Earth had been identified and mapped but much of it was in wilderness. Since the distance between two places in civilized Europe was too short for Halley's proposal, adventurers set off for destinations from Siberia to India and Sumatra. More than a dozen adventurers participated in the transits of 1761 and 1769, but, as we shall see, not all were successful. At that time, science had not advanced to the point where some scientists studied astronomy, others chemistry, some others physics, and so on. They were all called "natural philosophers," and one might travel to a distant land and come back describing the plant and animal life, the type of people, the geography, and the appearance of the heavens. So, these adventurers set off to do more than time the transit of Venus. They planned to provide detailed observations of the places they visited.

How did they pay for this? Now, in the twenty-first century, scientific endeavors are financed by charitable foundations or governments, but this type of support has only been widely available since the end of World War II in 1945. In the eighteenth century, the scientists were either independently wealthy or were supported by wealthy patrons. Jean-Baptiste Chappe d'Auteroche and Guillaume le Gentil were minor noblemen in France. Christian Mayer and Andras Lexell were bankrolled by the Russian Empress Catherine the Great. Charles Mason and Jeremiah Dixon set out for Sumatra with funds raised by subscription in England.

Nor was there any kind of a central scientific archive. There were academies of science supported by the rulers of France, Russia, and Sweden. But, by far the most influential scientific body was the Royal Society of London. Correspondents from all over Europe sent letters to be read at Royal Society meetings. For instance, Anton Leeuwenhoek, the inventor of the microscope, sent detailed drawings of tiny things invisible to the naked eye from his home in the Netherlands. Adventurers sent letters describing their observations in far off newly discovered lands. The Reverend Thomas Bayes submitted his mathematical musings. The Royal Society decided to coordinate these various sightings of the transit of Venus and gather them together after the 1769 transit to determine the distance from the Earth to the Sun.

After the transit of 1761, the observations began to arrive. An even more concerted effort was made to time the transit in 1769, with adventurers traveling to St. Petersburg, Canada, Baja California, Tahiti (the famous Captain Cook was involved in this one), Norway, Philadelphia, and Manila.

These adventures were not without their dangers. In 1756, war broke out between England and France and their allies. In the United States, this is remembered as the French and Indian War, where George Washington learned how to make war. It was called the Seven Years War in Europe and lasted until 1763. For the 1761 transit, many of the adventurers carried letters from both English and French diplomats that would supposedly get them through battle lines. However, these letters were not always adequate.

When Charles Mason and Jeremiah Dixon set out from England for the 1761 transit, their ship was attacked by a French warship, killing 10 of the sailors and forcing their crippled vessel back into port. Guillaume le Gentil was traveling in the Indian Ocean when he learned that the British had taken the French colony of Pondicherry in India, where he had intended to take his measurements. The ship turned around and headed back to Mauritius, but le Gentil was at sea on a tossing ship on the day of the transit and did not have a stable platform to make his measurements. The observations for the 1769 transit were taken in a more peaceful world and were more complete.

Of all the adventurers, Guillaume le Gentil was singularly unlucky. He missed the 1761 transit because he was on board a ship in the Indian Ocean. He stayed around for the 1769 transit and decided to view it from the Philippines. The Spanish authorities in Manila suspected him and did not understand his mission, but he was able to set up his instrument and showed Spanish officials and their wives the wonders of the heavens in the clear night sky. Le Gentil spent 3 years in Manila preparing for the 1769 transit, but the French Academy of Science contacted him in 1768 and urged him to return to Pondicherry, which had just been recaptured by the French, in India.

The night before the transit, the sky was clear at Pondicherry, but the day of the transit, the weather got worse and worse, with clouds piling up and blocking any view of the Sun. (It was a clear day, perfect for viewing the Sun, in Manila.) He returned to France to find that, when he had not been heard from for so long, he had been declared dead. His wife remarried, and his relatives split up his fortune. But, all was not lost for le Gentil. He wrote the story of his adventures, and this memoir became a bestseller in Europe. In 1992, the Canadian writer, Maureen Hunter, turned le Gentil's travails into a play, and, in 2007, she wrote the libretto for an opera—both named "The Transit of Venus." The complete story of the adventurers who went out to time the transits can be found in References [2–4].

In the months that followed the 1769 transit, the final observations of these adventurers were sent to the Royal Society of London. The Royal Society appointed a committee to examine the numbers and use them to compute the distance from the Earth to the Sun. The chairman of that committee was Lord Henry Cavendish (Figure 1.2). The Cavendish family was involved in scientific ventures for at least two generations, but Henry Cavendish was the most distinguished, having made major contributions to astronomy, chemistry, biology, and physics. Henry Cavendish had investigated the nature of hydrogen, measured the density of the Earth, showed the nature of fermentation, and had played a major role in the movement to make science dependent upon very careful measurements. Remember, the "natural philosophers" of that time were not walled off into separate disciplines, so a genius like Henry Cavendish could make contributions to many fields.

The Cavendish committee kept careful minutes of their meetings, and those minutes and the original data they received are in the archives of the Royal Society. How were they to proceed?

FIGURE 1.2   Henry Cavendish (1731–1810) whose committee used the data from the observed transits of Venice to estimate the distance from the Earth to the Sun. (Courtesy of istock.com.)

(In 1977, Stephen Stigler used those records and applied modern computer-intensive statistical methods to see if he could improve on the conclusions of the Cavendish committee but more of that in Chapter 10.)

Edmund Halley had proposed that, if the transit of Venus were to be observed at two points on the Earth, the difference in timing and the distance from one point to the other could be used to compute the distance from the Earth to the Sun. Depending upon how many of the observations they found that met their standards, the Cavendish committee could have had as many as 45 pairs of observations available, although the accuracy would be greatest if the two comparable measurements were made from places on either side of the equator. Which pairs of observations should be used?

Until the rise of statistical models in the twentieth century, scientists would consider a collection of observations or experimental results like these and carefully choose the "best" ones or one. For instance, in the 1870s Albert Michelson attempted to measure the speed of light at the Naval Observatory in Washington, DC. His idea was to send a thin beam of white light down two different pathways of mirrors. He could adjust the mirror positions on one path until he had dark rings of interference on the screen where he projected both beams at the end. The difference in length of the two paths could be used to measure the speed of light.

To do this, Michelson needed a steady source of pure white light. None of the artificial light sources available at that time produced the type of beam he needed. However, there was a crack in the wall of the observatory. For a brief period during the day, the Sun shone directly through that crack. Thus, he had a brief moment each day to take his measurements. Although he recorded all his runs, he used only a small subset of these for his final analysis. This was because he trusted his experimental experience and sense to tell him which of the runs was "correct."

(Nowadays, when most science is sponsored by government or foundation grants, it is considered a scientific fraud to select the "correct" values from a set of observations or experiments. The reputable scientist is expected to display all the data. This book will show why and how this came about.)

And so, in the spirit of eighteenth- and nineteenth-century science, the Cavendish committee examined the data that had been collected, searching for the ones that were "correct."

1.1 ERRORS

The first problem they ran into was called the "black dot" problem. It turned out that the adventurers could not observe exactly when the black dot of Venus crossed onto the disk of the Sun and when it left. Instead, they saw the black dot stretched and enlarged as it approached the limb of the Sun's disk and then suddenly appeared within the disk, without their being able to observe the exact moment of crossing. A similar problem occurred as the black dot emerged from the Sun's disk. The calculations Halley had proposed assumed that the time of crossing could be determined down to the second.

This is a problem that occurs in almost all scientific observations. The most careful measurements produce different numbers when the observations are run more than once. This failure to be able to replicate exact measurements descends all the way down to atomic physics, where current theory holds that the position or spin or whatever we wish to measure on an individual particle is fundamentally random.

In this book, we will call these kinds of differences in measurement "errors." Section II of this book will show how modern statistical methods deal with "errors."

1.2 BLUNDERS

But, the Cavendish committee had to contend with more than just random error. One of the adventurer's numbers did not match anything else in their data. When they made his time of transit part of the equations, the result was dramatically different from their other calculations. They eventually concluded that he had gotten the longitude of his position wrong. It was, in fact, very difficult to measure the longitude of one's position on the Earth during the eighteenth century. The latitude is easily determined by the height of the Sun from the horizon at noon, but the longitude has to be determined by knowing the exact time of the observation at the place where you are and also at some fixed spot on the Earth (like the Greenwich meridian).

The failure to have the correct longitude was not an "error" like the black dot problem. In the 1920s, the statistician William Sealy Gosset made a distinction between "errors" and "blunders." Errors were the differences in measurement that are inherent in the act of measuring. Blunders are something else.

I was once involved in an investigation of recorded temperatures in a large fermentation reactor that was supposed to be kept at a constant temperature. Once an hour, to determine the current temperature, the operators had to pull a small bucket of material out of the reactor, measure its temperature, and adjust the cold water flow into the reactor accordingly. We examined the records of 1 week, where, at the beginning of a new shift, the temperature suddenly dropped and had to be adjusted back, and, at the beginning of the next shift, the temperature suddenly rose and had to be adjusted down. Someone was sent to observe what was happening on those two shifts.

Drawing a bucket of material from a reactor was a dirty job. The worker had to climb up metal stairs, reach around in a small space, and open the spigot. So, that job was usually given to the least senior man on the shift. It turned out that the man who drew the sample on the first shift had just been hired, handed the bucket and the thermometer and told to get the temperature of the fermentation tank. He was observed to climb up to the spigot, draw the sample of material, take the thermometer, shake it hard up and down, plunge it into the bucket, wait a few seconds, then take the thermometer out and carry it over to the light to read it.

The only thermometers he had experience with were the clinical fever thermometers that might have been used at his home. In a clinical thermometer, a small kink in the mercury column keeps the mercury at the maximum value it reached unless the thermometer is shaken down. He did not realize that his thermometer registered whatever was the ambient temperature, so his "measurement" of the temperature of the reactor was, in fact, a measurement of the ambient air around the reactor. In the words of Gosset, all of his measurements were "blunders."

Section III of this book deals with how blunders are identified and handled in statistical analyses.

1.3 LIES

One of the adventurers bothered the Cavendish committee. He was a man who had been known to exaggerate some of his "findings" in previous adventures. Could his data be trusted?

Scientific activities have been bedeviled by instances of fraud since the beginning of the modern age. In spite of the honest attempt by most explorers and scientists to uncover the true nature of the world, there are isolated examples of something else.

For an early example, consider the voyage of discovery of Sebastian Cabot in 1508. Cabot left Bristol, England, with two ships and three hundred men, financed by the king, Henry VII. He sailed across the North Atlantic, discovered Newfoundland and traveled into Hudson Bay, which he thought was the beginning of a northwest passage around the Americas to the Orient. But, his men revolted at traveling further, and he returned home. At least, that's the story he supposedly told of his voyage. But, there is no contemporary record of that voyage, and all the evidence that we have for it are the statements of those who heard it from Sebastian. On the strength of his tales, Sebastian Cabot was named Master Mariner by the King of Spain and given the authority to oversee all future voyages of discovery from Spain to the New World [5].

Though we have no confirmatory knowledge of his 1508 voyage, we know from contemporary critics that he was a blowhard and told different versions of the voyage to different people. We know that he led one Spanish expedition to South America in 1526, which he botched through his inept seamanship and his provocative treatment of the Indians, who ambushed and killed most of his men.

Fortunately, the Sebastian Cabots of science are few, but they are there, publishing data from experiments that were never run, displaying doctored photographs to "prove" their conjectures, and describing in painstaking detail the results of observations on pairs of twins who never existed.

At one time, the most a scientist gained from his or her studies was prestige. And, there were still some who falsified data to get that prestige. But, since the end of World War II, science has become a more lucrative "business." A successful scientist can expect to have her or his work supported by the government or by grants from private foundations. The lure of money and prestige sometimes proves too much for a less than honest scientist. How can you know when data have been cooked or falsified? Section IV of this book shows how statistical methods have been used to identify the liars.

So, after all of this, how far is it from the Earth to the Sun? Current estimates put the length of an AU at 92,955,807 miles. The estimate of the Cavendish committee was off by only 4%—not bad for an eighteenth-century committee that had to overcome the "black dot" problem to find the correct answer without the aid of modern statistical methods.

1.4 SUMMARY

In the eighteenth century, the Royal Society of London collected observations from adventurers who went to different places in the world in order to time the transit of Venus across the Sun, which occurred in 1761 and 1769. A committee was formed to examine these reports and use them to determine the distance from the Earth to the Sun. In the process, they had to contend with small uncertainties that resulted from the inability to determine the exact time of transit. These are called "errors" in modern statistical terms. One of the adventurers apparently did not determine his longitude correctly. In statistics, this is called a "blunder." One of the adventurers had been caught lying before. This raises the problem of detecting falsified data.

REFERENCES

For a method of estimating the Astronomical Unit from the transit times of Venus

1. http://profmattstrassler.com/articles-and-posts/relativity-space-astronomy-and-cosmology/transit-of-venus-and-the-distance-to-the-sun/ (last modified June 2012).

For the history of the transit of Venus expeditions, 1761, 1769

2. http://www.astronomy.ohiotate.edu/~pogge/Ast161/Unit4/venussun.html (last modified May 2011).
3. http://sunearthday.nasa.gov/2012/articles/ttt_75.php (last modified June 2012).

4. http://www.skyandtelescope.com/astronomy-news/observing-news/transits-of-venus-in-history-1631–1716/ (last modified June 2012).

For the story of Sebastian Cabot

5. Roberts, D. (1982) *Great Exploration Hoaxes*. San Francisco, CA: Sierra Club Books, pp. 24–39.

II

Errors

CHAPTER 2

Probability versus Likelihood

T HE MOST CAREFUL SCIENTIFIC investigations have always been bedeviled by uncertainty. Multiple observations of the same thing usually yield multiple values. Unexpected events intervene in field experiments. Patients get well or succumb without any clear-cut "cause." Until the beginning of the twentieth century, the scientist used his (once in a while her) judgment to pick those numbers that were most "correct." In the hands of a perceptive genius like Henry Cavendish or Albert Michelson, these carefully chosen subsets of the available data came very close to the truth.

But, in the hands of someone less gifted, "results" were announced that later turned out to be not quite correct or completely wrong. This failure of judgment can be seen in the history of medical science. As George Washington lay in bed with a fever and crippling cough, his physicians bled him again and again to bring down the fever and hastened his death. In 1828, Pierre-Charles-Alexander Louis (1787–1872) decided to check on the efficacy of bleeding. He compared the records of patients with fever who had been bled and those who had not. The average time to recovery for the unbled patients was almost half the average time to recovery for those who had been bled. The year after he published his results, the number of leeches imported into France (for bleeding) increased by over 20%. The medical community of France was not yet ready for medical knowledge based on careful counting rather than the word of ancient philosophers like Galen, who had recommended bleeding as a cure for fever.

Because the "expert" judgment of the scientist had proven to be wrong so often, statistical models came to be adopted by more and more fields of science, starting in the last years of the nineteenth century with biology. By the 1930s, statistical models were being used in physics, chemistry, sociology, psychology, astronomy, and economics. Medical science was one of the last fields to begin using statistical models in the late 1950s and early 1960s.

The basic idea of statistical models can be summed up in Equation 2.1:

$$\text{Observation} = \text{truth} + \text{error} \qquad (2.1)$$

Suppose, we wish to measure the angular distance between two planets in the sky. Our measurements never quite get the "right" answer because of atmospheric turbulence, slippage in the gearing of our telescope, or an intervening cloud. These deviations from the "truth" can be thought of as coming from a collection of possible errors. Those errors have been described as a cloud of uncertainty. The physicist George Gamow (1904–1968) described them in terms of an uncertain billiard ball. At first glance, the billiard table appears to be filled with shifting melding balls, all of them "vague and gruely," but there is only one ball. The level of uncertainty (called Planck's constant in Gamow's world of atomic physics) is several thousand times greater for this ball than for objects we normally see.

The elements of this cloud of uncertainty (the set of all possible errors) can be described in terms of probability. The center of the cloud is the number zero, and elements of the cloud that are close to zero are more probable than elements that are far away from that center. We can be more precise in this definition by defining the cloud of uncertainty in terms of a mathematical function, called the probability distribution. There are a great many formulations of probability distributions, and some of them can become quite complicated. For the most part, in this book, I will restrict the discussion to two specific distributions, the normal distribution (sometimes referred to as the "bell-shaped curve") and the Poisson distribution.

Now, the poor scientist who has taken different measurements of the same thing has to deal, not only with different measurements, but a theoretical entire set of possible errors. How does this bring him closer to the "truth?" Why is this any better than the careful selection of data that the Cavendish committee used to determine the distance from the Earth to the Sun? To answer this, let's look at a relatively simple probability distribution that deals with the number of raisins in a slice of cake.

Raisins have been introduced into the cake batter, and the batter is stirred. Any one raisin does not know where the other raisins are, and the batter is not a smooth emulsion of crushed raisins, so there is a chance that in any particular slice of the cake we get no raisins, two raisins, or three. There is even a very small probability that our slice has all the raisins. The probability of getting at least one raisin in the slice depends upon how thick is the slice. The thicker the slice, the greater is the probability that we will have at least one raisin in it.

Israel Finkelstein (b. 1949) of Tel Aviv University is one of the leaders in the use of statistical methods in archeology. To illustrate how we can extend this idea of raisins in a cake to more complicated situations, consider a survey he organized hunting for evidence of human settlements in the West Bank of the Jordan River during the twelfth and eleventh centuries before the common era (b.c.e.) (see Reference [1]). Like the slice of cake, any given area might have evidence of zero, one, or two habitations from that time. If these homes were established at random, the probability of finding one or more homes in a given area would follow the same statistical pattern. Finkelstein was interested in finding evidence that these homes were clustered in some way, representing different tribal groups. He was looking for evidence that his "raisins" did know about other "raisins" and clustered together.

For Finkelstein, the "slice" was an area of land. Other archeologists who dig pits in formerly inhabited caves are looking for evidence of human habitation like stone spearheads in a region between two well-dated layers of debris. Their "slice" is a three-dimensional region. We call any region or grouping of things in which we expect to find specific events a "substrate." This word is taken from the idea of an archeologist or a paleoanthropologist isolating a region in a dig under or defined by some geological structure, like a layer of volcanic ash.

In a general fashion, then, we have a mathematical model where we are trying to count some things scattered more or less evenly through some substrate in such a way that the probability of seeing at least one event is proportional to the size of the section of the substrate we are examining. This could be the number of pregnant women who suffer miscarriages, the number of cosmic rays that penetrate a detector over a fixed period of time, or the number of industrial accidents that occur in a particular plant in a given month. This type of a probability distribution is called a Poisson distribution named after the French mathematician Simeon Denis Poisson (1781–1840). I have chosen it for this book because it has a relatively simple mathematical structure, and I can discuss fairly sophisticated ideas without getting involved in a great deal of mathematical notation.

The Poisson distribution states (in algebraic notation) that the probability we will observe x events equals

$$\frac{e^{-\theta}\theta^{x}}{x!} \tag{2.2}$$

This is not very different from the use of algebraic notation in high school algebra. In high school algebra, for instance, we represent the relationship between distance, rate, and time with the formula

$$d = rt$$

As Sir Isaac Newton pointed out, this enables us to determine the value of any of these three (d= distance, r= rate, t= time) from the knowledge of the other two.

When we use algebraic notation in statistical models, the problem becomes more complicated because we cannot "observe" a probability and know its exact number. We can only estimate probabilities on the basis of observations. However, the same principle applies. We use letters to stand for numbers in the general formula and plug in what numbers we can observe to "solve" for the numbers we cannot observe.

There are three letters that stand for numbers in formula (2.2), **e**, **θ**, and **x**. The number e is a fixed constant, like the number π (the ratio of the circumference to the diameter of every circle). The number e is derived in the supplement to the chapter for those who want to see the exact mathematics behind it. But, like π, it has a fixed value that is always the same. The letter x stands for a number we can observe. In the case of the Poisson distribution, it is the number of raisins we found in a specific slice of the cake. The third number is represented by the Greek letter, theta, θ. This is a number which governs the exact values of the Poisson probability that can be used to describe the problem at hand.

Numbers that govern the shape of the distribution but cannot be observed are called **parameters** and, by convention, they are almost always represented by Greek letters.

Using the Poisson distribution, the probability of observing seven raisins in our slice of cake is

$$\mathrm{Prob}\{x = 7\} = \frac{e^{-\theta}\theta^7}{7!} \tag{2.3}$$

(Recall from algebra that the symbol $7! = 7 \times 6 \times 5 \times 4 \times 3 \times 2 \times 1$.)

The idea behind statistical methods is to use the observed numbers (the Roman letters) to estimate the values of the parameters (the Greek letters).

Notice the use of the word "estimate." In mathematics, we often use common everyday words, but we assign precise meanings to those words. In ordinary language, we can often use words with vague or multiple possible meanings—in mathematics everything has to be carefully defined. An estimate (the mathematical definition) is a number derived from observed values that is as close as we can get to the true parameter value. Useful estimators are those that are "better" in some sense than any others.

(A friend of mine was called as an expert witness in a lawsuit, and the attorney for the other side said to him, "Dr. So-and-So, isn't it true that what you have presented here are nothing but estimates?" The attorney was confusing the exact mathematical meaning of "estimate" with its vague everyday meaning as some sort of guess.)

What properties should a good statistical estimator have? Since we are dealing with probability, we start with the probability that our estimate will be very close to the true value of the parameter. We want that probability to become greater and greater as we get more and more data. This property is called **consistency**. This is a statement about probability. It does not say that we are sure to get the right answer. It says that it is highly probable that we will be close to the right answer.

The previous paragraph deals with the center of the distribution of uncertainty associated with the estimator. However, we need to look at another aspect of that cloud. Two clouds of uncertainty may have the same center, but one may be much more dispersed than the other. We need a way of looking at the scatter about the center. We need a measure of the scatter. One such measure is the variance. We take each of the possible values of error and calculate the squared difference between that value and the center of the distribution. The mean of those squared differences is the **variance**. There are other measures of the scatter, but, in this book, it will be sufficient to use the variance as a measure of the scatter.

Now, back to the question of what makes a good estimator of a parameter.

A good estimator has to be more than just consistent. It also should be one whose variance is less than that of any other estimator. This property is called **minimum variance**. This means that if we run the experiment several times, the "answers" we get will be closer to one another than "answers" based on some other estimator.

The genius who founded much of modern statistical theory, R.A. Fisher (1890–1962), proved that these two properties, consistency and minimum variance, are true if we calculate the estimator in the following way:

Equation (2.3) gives the probability that our observation, x, equals 7. If we have actually found seven raisins in our slice of cake, then this makes no sense. The probability that there are seven raisins in the slice is 1.0, for we actually observed that. So, Equation (2.3) is not a probability. Fisher called it a **likelihood**. It is a formula that connects the unknown parameter, θ, with the observed value, 7. Fisher proved that the values of the parameters that maximize the likelihood have the two properties we seek, consistency and minimum variance, and so, in most cases, are better than any other method of estimating the parameters. These are called the **maximum likelihood estimators**. In Equation (2.2), $\theta = 7$ maximizes the likelihood, so our best procedure is to use our one observation. But, usually, we get more than one observation. What does the likelihood look like then?

Industrial accidents in a large plant occur from time to time, and workers are often injured or even killed in such accidents. Suppose a company has introduced a new education program designed to decrease industrial accidents. How can they know if the new program is actually "working?"

In the 1920s, psychologists thought that they had discovered a major source of industrial accidents. They noted that some workers had been involved in multiple accidents, and they called such workers, "accident prone." For the next 50 years, efforts were made to identify the characteristics of "accident prone" workers and see what made them different. All of these attempts failed. This is because there is no such thing as an "accident prone" worker. Industrial accidents have been shown to follow a Poisson distribution of statistically independent values, and the probability that two different workers will be involved in accidents is equal to the probability that any one worker will be involved in two accidents. Reference [2] describes the vain search for "accident proneness."

Getting back to the company that has introduced a new education program, suppose we collect the accident data for the 6 months prior to the new program and for the 6 months after the new program. And, suppose the number of accidents in each of the prior 6 months are

$$6, 2, 5, 0, 7, 3$$

Using the formula for six independent Poisson variables, the likelihood of that sequence is

$$\frac{e^{-\theta}\theta^{(6+2+5+0+7+3)}}{(6!2!5!0!7!3!)} \tag{2.4}$$

Suppose that after the introduction of the new education program, the next 6 months accidents are

$$0, 2, 0, 5, 12, 1$$

The likelihood of that sequence is

$$\frac{e^{-\theta}\theta^{(0+2+0+5+12+1)}}{(0!2!0!5!12!1!)} \tag{2.5}$$

If we were back in the early nineteenth century, when scientists used the data that were closest to being "correct," we might find two opposing views. One would say that the program worked since all but two of the post education months had very few accidents. The month with 12 accidents was obviously different, and something unusual must have happened, so we can ignore it. Another group might say that it looks like the program, if it worked, worked only for the first 3 months.

In this one-parameter model, it turns out that the estimator of the parameter θ that maximizes the likelihood is the average of the observations, and the average number of accidents/month before the new education program was 4.333, while the average after the new program was 3.333.

Is there really a difference between the two sequences? Remember that our estimates consist of the truth plus an error. We have chosen estimators that minimize the uncertainty associated with the error, but have we minimized it enough to say whether 4.333 estimates a different parameter than 3.333?

As the reader can now see, once you admit statistical models into your science, it is a little like stepping on a piece of gum. The more you walk on it, the more it gets involved in every step you take. We will stop here and leave that question to a formal course in statistics.

2.1 SUMMARY

The basic statistical model is that

$$\text{Observation} = \text{truth} + \text{error}$$

where "error" is defined in terms of a probability distribution. Each probability distribution can be described as a mathematical formula where, by convention, Roman letters are used to represent observations and Greek letters are used to represent specific characteristics of the distribution, called parameters.

Estimators are functions of the observed values that can be used to estimate specific parameters. Good estimators are those that are consistent and have minimum variance. These properties are guaranteed if the estimator maximizes the likelihood of the observations. This is illustrated for the Poisson distribution, which has only one parameter.

2.2 FOR THOSE WHO WANT TO SEE THE MATHEMATICS

The formula for the Poisson distribution contains the number denoted as "e." There are two ways of calculating the value of e. We start with the formula for compound interest from finance. If we take one dollar and add a small percentage, p, each year, then at the end of 10 years, that dollar will be equal to

$$(1+p)^{10}$$

Suppose, we decrease the amount of interest while we increase the number of periods during which the interest is paid. Let x = the number of periods, and let $1/x$ = the interest. Then, the formula becomes

$$\left(1+\left(\frac{1}{x}\right)\right)^{x}$$

What happens if x increases indefinitely? The interest, $1/x$, gets smaller and smaller, so the form inside the parentheses gets closer and closer to 1.0 and raising 1.0 to any power, no matter how large, does not change it. However, as we increase the value of x, $1/x$ is not quite zero, so we are raising a number just slightly bigger than 1.0 to the power x. These two contrary tendencies coalesce into a number

2.718281828459045235360287471352662497757247093699959574…

This displays the first 54 values in the decimal expansion of e, but the decimal expansion goes on forever, and computer programs have been used to determine the first 10 million of those digits.

Euler showed that there is another way to determine the value of e [3]:

$$e = 1 + \frac{1}{2!} + \frac{1}{3!} + \frac{1}{4!} + \frac{1}{5!} + \ldots (\text{forever})$$

The perspicacious reader will see how the formula for the Poisson probability distribution arises from this last formula.

REFERENCES

1. Finkelstein, I., and Lederman, Z. (1997) *Highlands of Many Cultures, The Southern Samaria Survey.* Jerusalem: Graphit Press.
2. For the history of "accident proneness," see http://www.ncbi.nlm.nih.gov/pmc/articles/PMC1038287/ (last modified Jan 1964).
3. http://www-history.mcs.st-and.ac.uk/HistTopics/e.html (last modified June 2002).

CHAPTER 3

The Central Limit Conjecture

SIMON-PIERRE, MARQUIS DE LAPLACE (Figure 3.1), set himself the task of using Newton's laws of motion to derive the exact orbits of the known planets. For his magnificent tour de force, named *Mécanique Celéste,* he collected all the observations he could find of relative positions in the sky among the planets and the sun. He tried to use them in the detailed mathematical derivations dictated by Newton's laws. All the observations? Now, that posed a problem. All these historical observations, so carefully conducted by previous and contemporary astronomers, contained errors.

In 1738, a few years before Laplace was born, Abraham de Moivre (1667–1754) had shown that, if you add up a lot of small random numbers of a certain kind, the probability distribution of their average could be represented by the formula

$$\left(\frac{1}{\sqrt{(\pi\sigma^2)}} \right) \exp\left\{ -\left(\frac{x-\mu}{2\sigma} \right)^2 \right\}$$

FIGURE 3.1   Simon-Pierre Laplace (1749–1827), a French mathematician, who used Newton's laws of motion to determine the exact paths of the known planets. (Courtesy of Shutterstock.com.)

Here, π is the transcendental number that describes the ratio of the circumference of a circle to its diameter, and the notation "exp{ }" means the transcendental number e is raised to the power of the expression inside the parentheses.

Laplace realized that it was very difficult to get exact measurements of relative positions of planets with the instruments of his time and earlier, so he conceived of the observations being modified by the sum of a lot of little problems, like the shimmering of the air, the gearing on the transit, and so on. He added De Moivre's probability distribution onto his equations, and he called it the "error function."

This error function has two parameters, represented by the Greek letters μ and σ. The first, μ, describes the center of the probability distribution (which, in Laplace's formulation was set at zero). The second, σ, describes how far most of the error values are scattered away from μ. We call μ the **mean** and σ the **standard deviation**. The square of the standard deviation, $σ^2$, is the **variance**.

Just as we use Greek letters for parameters (which cannot be observed but only estimated) and Roman letters for observed values, we need to keep the distinction between the probability distribution and its parameters and the observations and the estimates derived from them. Thus, we call the parameter, μ, the mean, and we call the average of the observations (which are often used to estimate the mean) the **average**. When it comes to the second parameter, σ, the terms for the theoretical parameters are **standard deviation** and **variance**. The terms for the estimators derived from the data are **sample standard deviation** and **sample variance**. In the applied statistics literature, this distinction is not always adhered to, leading to unnecessary confusion.

Through the rest of the nineteenth century, Laplace's error function was used by others who wanted to introduce error probabilities into their calculations. What Laplace called the error function has been called the Gaussian distribution, the **normal distribution**, and the "bell-shaped curve." As statistical models began to appear in the various sciences, more and more workers made use of this normal distribution to describe the probability distribution of the error. They all assumed that one could represent the random errors as the sum of a lot of little errors and that de Moivre's probability distribution was appropriate. This was called the **central limit theorem**. I prefer to call it the "central limit conjecture," since it had not yet been proven.

The central limit conjecture had two great advantages. Before we had the modern computer, it was relatively easy to calculate the maximum likelihood estimators of μ and $σ^2$. It was tedious, often involving many pulls on the crank of a mechanical calculator, but it could be done. For the maximum likelihood estimators of these two parameters, the fact that the error was normally distributed meant that the estimator of μ was normally distributed and the estimator of σ² had an easily derived distribution. Furthermore, if you had N observations, the variance of the maximum likelihood estimator of μ was σ²/N. This meant that more observations resulted in smaller scatter in the cloud of uncertainty.

There are three properties of the normal distribution that make it attractive to use:

1. The average of observations has a normal distribution with mean equal to μ. (It is **unbiased**.)

2. The distribution of errors is **symmetric**, that is, positive errors are as probable as negative errors of the same size.

3. The variance of the average decreases at a rate of 1/(number of observations).

I have a friend who worked for the local state government, and he once posed the following real-life problem.

The state's Department of Weights and Measures regularly purchases products on grocery shelves and checks on the weights of the contents. Regulations require that boxes labeled as having 12 ounces, for instance, should have an average weight of 12 ounces but could vary by plus or minus 1.5 ounces. The range of variation was established years before and was based upon how accurate the filling machines could be. In the problem he posed, the department has purchased three 12-ounce boxes of parboiled rice made by company X, and all three boxes have exactly 11 ounces of rice in them (within the allowed range). However, the lack of variation and the underweight suggests that this company's filling machinery is accurate enough to control the weight much better than before and that the company is purposely shortchanging its customers.

The questions he posed were, how can the state determine if this is true? and how many boxes must be weighed to be reasonably sure of the conclusions? He noted that if this ever came to a trial, the number of boxes examined and the method used to show that there was a consistent underweight would have to be explained to a judge, a judge who might not have taken a course in statistics.

Here is how I would explain it to the judge. We start with the statistical mathematical model:

Observed weight of a box = intended average weight + error   (3.1)

We would like to use the average of the observed weights of N boxes to estimate the intended average weight, where N is a number determined by examining the statistical model. The statistical model proposes that the errors, the differences between the observed weights and the intended weights, come from a normal distribution. How many boxes have to be examined to be reasonably sure that company X is cheating its customers? The variance gives us the answer.

Suppose we look at the average of four boxes. The basic statistical model still holds:

Average of four boxes = intended average weight + error

But, the cloud of error is smaller than it was for one box. It still has a mean of zero, but its variance equals the variance of the original cloud divided by four. If we use the average of 15 boxes, the variance of the error is now the original variance divided by 15. With a simple calculation, you can find a sample size, such that we know that the average weight of our sample has so little error that we can be almost certain that the intended weight was very close to our average.

There is a constant battle between the cold abstract absolutes of pure mathematics and, the sometimes sloppy way in which mathematical methods are applied in science. What right do we have to assume that the central limit conjecture holds in this case?

In 1922, Jarl Waldemar Lindeberg (1876–1932) of Finland found a way to analyze the mathematical foundations of that conjecture. In the 1930s, Paul Levy (1886–1971) of France extended Lindeberg's work to establish the conditions necessary for the central limit theorem to hold. However, it remained very difficult to prove that a set of data meet the Lindeberg–Levy conditions. Then, in 1948, Wassily Hoeffding (1914–1991), a Finnish mathematician at the University of North Carolina, showed that the Lindeberg–Levy conditions hold for a large class of statistical procedures. As a result, most of the scientific papers that invoke the central limit theorem do so by showing that their methods use one of Hoeffding's class of procedures.

The normal distribution is not the only well-behaved cloud of uncertainty. Although these others are often not as easy to deal with, they have been found to be very useful. For example, the Nobel prize-winning economist Eugene Fama (b. 1939) has looked at the closing prices of stocks on a stock exchange. Who would not want a sophisticated statistical method for choosing the right stocks? He found that they do not fit the normal distribution but fit to a class of mathematical functions known as the "symmetric stable distributions," and you need advanced calculus to deal with these. (See References [1] and [2].)

When we say that another error distribution is "more difficult to deal with," it means that it will take many complicated mathematical calculations to estimate the effects of error on our observations. This was a valid objection when an analyst had only a desk calculator (or, before that, pencil and paper) to help with the calculations. Now, in the computer, we have an indefatigable servant, which cannot engage in original thinking (so far) but will run millions of calculations swiftly and without complaint. So, it is possible to introduce other error distributions into our calculations. It is possible to do so, and it is being done at research universities. However, the commercial statistical software that is used in many applications of statistical modeling is often based on the normal distribution.

In the following chapters, we will look at some examples where the simple model:

$$\text{Observation} = \text{truth} + \text{error}$$

and the fact that error has a normal distribution has been used to solve problems.

3.1 SUMMARY

The central limit conjecture states that most errors are the result of many small errors and, as such, have a normal distribution. The assumption of a normal distribution for error has many advantages and has often been made in applications of statistical models. In the 1930s, the conditions necessary for the central limit conjecture to hold were established and further refined in the 1940s.

REFERENCES

1. Fama, E.F., and MacBeth, E.d. (1973) Risk, return, and equilibrium, *J. Polit. Econ.*, 81, 607–636.
2. Fama, E.F., and Roll, R. (1971) Parameter estimates of symmetric stable distributions, *J. Am. Stat. Assoc.*, 66(334), 331–338.

CHAPTER 4

Measuring Disease

H ow do you measure a disease? In cases of an acute disease, like meningitis or strep throat, the doctor can determine if and when the patient recovers from the disease. But, what about a chronic disease, like heart failure or diabetes, where the patient's sickness increases or decreases, often without seeming to respond to any specific medication? How do you determine when the treatment is "working?" How do you measure patient response?

The development of modern science rested on careful measurement. Boyle's law resulted from the ability to measure both pressure and temperature of a gas. Newton's laws of motion assumed that speed (and, with it, acceleration) could be measured and presented as specific numbers. But medicine lagged behind. The doctor treats a patient with tuberculosis, but how does one measure whether the patient is improving or deteriorating? And so, the late nineteenth century and the early twentieth century saw the attempt by medicine to find ways of measuring disease.

Consider Chronic Obstructive Pulmonary Disease (COPD), where the lungs have lost some of their resiliency, and the patient finds it difficult to breathe in and out. COPD is usually a disease of smokers and involves permanent damage to the lungs that continues long after the patient has stopped smoking.

One way developed for measuring COPD is to have the patient blow as hard as he or she can into a tube, where the air is evacuated as it flows in, so there is no air pressure pushing back. The technicians who measure this (using a spirometer) are trained to encourage the patient to "blow, blow hard. Keep blowing! Keep blowing!" until all the air that the patient can breathe has been forced out. The total amount of air that is blown out is called the forced vital capacity (FVC) [1]. It ranges between 3 and 5 liters. However, there are differences due to gender, age, and height. Men tend to have higher FVCs than women. Taller people tend to have higher FVCs than shorter people. And, as you age, your FVC tends to be reduced. There are other factors that seem to have an effect on the FVC, though not as great as gender, age, and height. These include the person's ethnic background, normal exercise regimen, body fat, and even language skills. I was once involved in a study where one patient, who had a poor grasp of English, thought that when the technician shouted, "Keep blowing," the technician meant for the patient to stop.

So, how does the physician use FVC to measure a patient's COPD? It is compared to what would be "normal" for somebody like the patient. But, how do you determine what is "normal"?

One way to determine what is "normal" is to assemble a large number of people, who do not have any lung disease, and measure their FVCs. There is a certain range of values one might expect in healthy adult humans. You do not expect to find someone with 0.5 liter or someone with 20 liters. We could model the FVCs we measure from several thousand people as

$$\text{Observed FVC} = \text{overall adult mean} + \text{error} \qquad (4.1)$$

We could, then, use the average of all the observations to estimate the overall adult mean of all humans, and use the difference between that and each observation to model the error. We could use the normal distribution assumption to find values that will cover 95% of the population, and call these the "normal range." This is done, for instance, when blood chemistries are measured on a patient. One marker of kidney disease is the blood urea (BUN), and its normal range is 7–20 mg/liters. So, physicians observing a patient with kidney disease will try to drive his or her BUN down to this normal range.

However, if we attempt this for COPD, the "normal range" is much too wide. A young 6-foot male with the beginning of COPD might be able to force out more than 2 liters, while a 65-year-old female without COPD who is 5 foot 2 inches tall may have the same FVC.

We need to use a more complicated model:

$$\text{Observed FVC} = \text{overall adult mean} + \text{something about age} \atop + \text{something about height} + \text{error} \qquad (4.2)$$

And, we do this for males and females separately.

Imagine the corner of a room. Running along one wall at the floor, we mark out possible values of age. Running along the other wall at the floor, we mark out possible values of height. Using the data from one of the healthy volunteers, we find his age along one wall and his height on the other wall, and we plot the point on the floor that represents his age and height. We measure up vertically from that point to a level that represents his FVC, and we put a little blue light there. We go to another volunteer, find his age and height and go up to his FVC and put another blue light there. We do this with all the male volunteers, so our room now has a scattering of little blue lights. We notice that, as we go along the wall for age, the lights tend to go downward, and as we go along the wall for height, they tend to go upward. To anchor this collection of blue lights, we compute the average age, the average height, and the average FVC of all of the male volunteers, and we place a little red light there (see Figure 4.1).

Now, we bring in a large flat board. We pivot it at the red light of the average and twist it around until most of the blue lights are not

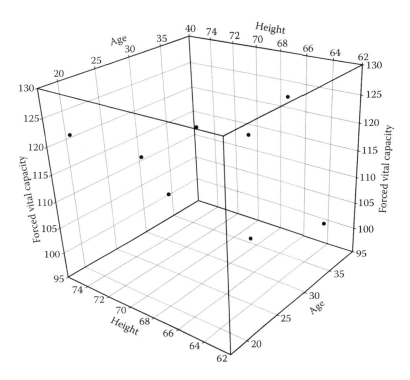

FIGURE 4.1   Three-dimensional plot of FVC measurements versus subjects' height and age.

too far from it. (There is a complicated mathematical procedure involving vector analysis and matrix theory that enables us to do this in a way that makes the various errors as small as possible.)

We do this separately for males and females.

The FVC values that do not quite fit on the board are very close to the predicted values of our board. In fact, we have used the data to make them as small as possible. We can use this board to predict the FVC of a patient if that patient were "normal." The measure of the patient's COPD is now

100× (patient's observed FVC)/(predicted FVC) = percent of normal

Notice what our model does not do. It does not tell us that height or age "cause" the FVC, whatever that might mean. It only tells us how to best predict FVC from height and age. It does not tell us whether a patient with FVC lower than predicted has COPD. It only tells us how severe that patient's disease is after a proper diagnosis of COPD. We will meet the confusion that results from identifying elements of a statistical model as showing that something "causes" the result as we examine other statistical models in this book. This confusion between "cause" and statistical correlation will be addressed in Chapter 7, and we will see that much of the confusion results from the fact that "cause and effect" are ill-defined concepts, whereas correlation is something that is a specific part of a statistical model.

This use of a flat board to estimate the relationship between FVC and height and age is called a "multilinear model," since the mathematical formulation that describes that board belongs to a class of models that are based on the mathematical formula for a straight line in geometry. If we use a multilinear model to predict FVC from heights and ages that are within the range of the volunteers' heights and ages, then the value predicted by this multilinear model is very close to the value that might have been predicted by a more complicated model. Because this is so, multilinear models are used in all branches of science.

In normal conversations, we don't have to be careful to translate our ideas into mathematical formulas. In such conversations, you will sometimes hear someone (who is attempting to show that he or she can speak science talk) say, "This problem is nonlinear." That person really means that the problem is complicated. But, the difference between linear and nonlinear models is not the same as the difference between simple and complicated. We shall see in future chapters how specific characteristics of the data require us to use specific nonlinear models.

One final warning about the use of statistical models (whether linear or otherwise): The estimated model describes the structure of the data that have been observed. It is unwise to extend this model very far beyond the observed data. Having observed FVC values from several thousand healthy people, we can predict the normal FVC for someone like them, but we don't know the "normal" FVC for a 95-year-old 7-foot 3-inch former basketball player—unless we had several people like that among our volunteers.

4.1 SUMMARY

In order to evaluate measurements taken on a patient, the doctor needs to know what values of those measurements are "normal." In many cases (e.g., blood chemistries) it is sufficient to know the average value for healthy people and use the normal distribution for error to determine the "normal range"—the values that 95% of healthy patients will have. In more complicated situations, we can use a more complicated model that takes into account aspects of the patient that affect that measurement.

An example is forced vital capacity (FVC), used to measure COPD. The "normal" value is a function of patient's gender, age, and height, among other things that are less influential. The statistical model is

Observed FVC = (overall mean for that gender)
+ (something involving age)
+ (something involving height) + error

where this is a multilinear model, it can be envisioned as a large flat board in three-dimensional space. Such multilinear models are widely used in scientific investigations.

4.2 FOR THOSE WHO WANT TO SEE THE MATHEMATICS

The multilinear model described in this chapter can be written as

$$y = \beta_0 + \beta_1 x_1 + \beta_2 x_2 + \text{error} \tag{4.3}$$

Let's look at this formula (4.3) and see why it is written the way it is. The basic idea behind using algebraic formulas in science is to use the rules of algebra (and logic) to find answers to questions. Off to one side, we write what the various symbols stand for

$$y = \text{FVC}$$
$$x_1 = \text{age}$$
$$x_2 = \text{height}$$

As we go about manipulating the algebra, we know that we can always go back and see what these symbols stand for, but, otherwise, we can forget their meaning and just run the algebra to see what happens.

REFERENCE

1. National Tuberculosis Association. (1966) *Chronic Obstructive Pulmonary Disease, a Manual for Physicians.* National Tuberculosis Association, Portland, OR, p. 31.

CHAPTER 5

Other Uses of Multilinear Models

IN 1977, THE U.S. Supreme Court ruled in favor of the government in an antitrust case against the Container Corporation of America. The government's contention was that Container Corporation and other manufacturers of cardboard containers and corrugated sheets had engaged in a conspiracy in restraint of trade to fix prices for their products. The companies convicted of price-fixing contributed to a fund of $300 million to compensate the victims of this fraud, for which the law allowed them triple damages [1].

But, what did "triple damages" mean? How much had the purchasers of boxboard lost? What was the difference between what they would have paid if competition had been allowed and what they did pay during the period of the conspiracy (1962–1975)? The price of cardboard boxes fluctuates from month to month depending upon the cost of raw goods, the price of the previous month, changes in cost of production, and the general average level of demand. There are also minor effects due to severe weather, transportation costs, and whether production runs were short or long.

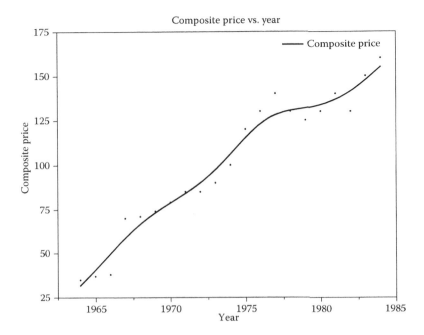

FIGURE 5.1 Yearly averages of a composite price of boxboard from 1964 to 1985. Was there an effect due to the conspiracy to fix prices from 1965 to 1975?

Figure 5.1 shows the yearly average of a composite price of boxboard during 1963–1985. It shows a general increase in price throughout the period of the conspiracy, which appears to continue at the same rate after 1975, when competition supposedly returned. Just looking at the graph, there is no clear indication that the conspiracy had any effect on price. But, there is a statistical model that can uncover the effects of the conspiracy.

Recall that in multilinear regression, we have modified the equation

$$\text{Observation} = \text{truth} + \text{error}$$

to define "truth" as a function of several inputs, so

$$\text{Observation} = (\text{multilinear model}) + \text{error}$$

where the multilinear model includes characteristics of the observation that are expected to influence the measurement.

One of the methods used to determine damages in the case of the Container Corporation of America was a multilinear regression of monthly composite price versus factors expected to affect the fair price of the product.

Price of goods for a given month = (overall mean for that month)
+ (price of the product during the previous month)
+ (change in costs of production for that month)
+ (level of output in industries using that product for that month)
+ (wholesale price index for that month)
+ (productive capacity of the cardboard box industry for that month)
+ (a "dummy variable" for the existence of the conspiracy)
+ error

This is an example of an econometric model. The second and third elements in the model (price during the previous month and change in costs of production) are variables that describe the general state of that industry at that time. The next two (output of industries using the product and wholesale price index) are measurements of the demand for the product during that time. The sixth one (productive capacity of the cardboard box industry) is a measure of supply for that period.

In economic theory, the price of goods in a free market should go up with demand and down with supply. But, supply and demand are not sufficient to describe the price of something. There are also underlying characteristics of the goods being sold. So, the first three items in the model describe the underlying characteristics of the goods being sold. The fourth and fifth items describe demand, and the sixth describes supply.

As the Supreme Court determined, the market for boxboard during the period of the conspiracy was not a free and open market. The price of goods was distorted by that conspiracy. And so, the model includes an additional variable that is equal to one during the period of the conspiracy (1963–1975) and equal to zero during other years. This is a **dummy variable**, a widely used element in statistical models. One often wants to know if some intervention in the process being examined had an effect. The estimated effect of the dummy variable indicated that the customers had been overcharged by 7.8% during the time of the conspiracy.

Dummy variables are used in any situation where we wish to determine the effect of a one-time intervention, in the face of a number of other variables that are correlated with the outcome. Here are some other examples.

The levels of air pollution in Los Angeles, California, from 1958 to 1975 were examined to determine the effect of new stringent automobile pollution regulations on automobile exhausts. In sociology, dummy variables are used to determine if some disturbing factor has had an effect. For example, one study examined the relationship between education level and the age of marriage for Japanese women and used dummy variables to represent different levels of education [4].

Dummy variables are used in epidemiology (what effect did the Chernobyl nuclear accident have on the incidence of leukemia in Europe?). They are used in demography (did the siege of Paris in the Franco-Prussian War have an effect on the steady growth of population in that city?).

The use of dummy variables shows that linear models need not deal with just the correlation of outcome versus measurable possible inputs. Linear models are not the opposite of "complicated." When using multilinear regression, we can be quite sophisticated.

Fisher arrived at Rothampsted Agricultural Station (north of London) in 1920. He was confronted with 90 years of data collected from different fields on which different types of artificial fertilizer were used. Could he find the most useful combination of fertilizer from those data? In his first examination of the numbers, Fisher realized that the amount of rainfall was a far more important factor influencing the wheat output than any difference in types of fertilizer. To solve this problem, Fisher invented multilinear regression. But, the relationship between yield and rainfall was not a simple one. He found that he had to use the rainfall, the square of the rainfall, and the cube of the rainfall. He also had to introduce dummy variables to account for arbitrary changes in fertilization patterns. Users of multilinear regression have dealt with models more complicated than Fisher did, using input variables that rise and then level off, that have infinite singular points, and so on [2].

In fact (with one major exception), there are no limits on what kinds of variables can be used in a multilinear model. If we are looking for the effect of extreme values of some influential factor, we can use a variable that is equal to zero unless that factor is greater than some threshold number, and then equal to that variable. An example of this occurred with examination of the widespread animal carcinogens, the aflatoxins. These are produced by fungi that occur naturally wherever grains are stored. It is possible to detect their presence at levels as low as five parts per billion. Because of this, we can find aflatoxins in wheat, bread, corn and corn products, oats, and rye products. At very high doses, aflatoxins produce liver cancer in experimental animals. Levels approaching those used in the animal studies can be found in native African beers.

Dummy variables that were zero until the aflatoxin level was about 10 parts per million were used to determine the relationship between the incidence of liver cancer and exposure to aflatoxins from the native foods of Africa and Thailand. No clear effect on human cancer was found in these studies, but the subjects of the studies tended to die early from infections and trauma and not many lived long enough to get liver cancer (see Reference [3]).

There is one exception to the statement that the variables that are used in multilinear regression can be anything. If the variable in the regression is, itself, measured with statistical error, then the algorithm that is used in standard commercial statistical software to run multilinear regressions is not the best one possible. We need a different type of computation to get less variable (and unbiased) estimates of the parameters of a multilinear model when the "explanatory" variables are subject to random error.

5.1 SUMMARY

An example is shown of a legal case where multilinear regression was used to estimate the damages associated with a price-fixing conspiracy of artificially raised prices of cardboard boxes during the 1960s. This regression used a dummy variable to detect the effect of the conspiracy. A dummy variable is equal to one whenever the event being examined has occurred and zero when it has not. Additional examples of multilinear regressions, which used dummy variables, are described from epidemiology, sociology, and environmental studies.

REFERENCES

1. Finkelstein, M.O., and Levenbach, H. (1986) Regression estimates of damages in price-fixing cases, in DeGroot, M.H., Fienberg, S.E. and Kadane, J.B. (eds) *Statistics and the Law*. New York, NY: John Wiley & Sons, pp. 79–106.
2. Fisher, R.A. (1924) The influence of rainfall on the yield of wheat at Rothamsted, *Phil. Trans. B.*, 213, 89–142.
3. Gibb, H., Devleesschauwer, G., Bolger, P.M., Wu, F., Ezendam, J., Cliff, J., Zeilmaker, M., et al. (2015) World Health Organization estimates of the global and regional disease burden of four food-born chemical toxins, 2010: A data synthesis, *F1000Res.*, 4, 1303.
4. http://academicworks.cuny.edu/cgi/viewcontent.cgi?article=1007&context=gc_econ_wp (last modified Jan 2015).

CHAPTER 6

When Multilinear Regression Is Not Adequate

U P TO THIS POINT, we have looked at statistical models of the form:

$$\text{Observation} = (\text{multilinear model}) + \text{error}$$

and, in each of the examples described, there has been the implicit assumption that the probability distribution of the error is the normal distribution. There are times when the assumptions behind the use of the central limit conjecture are not true, and some other probability distribution has to be used to describe the error. Some of these can be quite complicated and are most easily explored using sophisticated programs on a modern computer.

In 1948, the National Institutes of Health began funding a study aimed at uncovering the predictors of heart disease. The study involved 5209 adult residents of the city of Framingham, Massachusetts. The subjects were given complete physicals, including a battery of lab tests, and they filled out a questionnaire regarding their lifestyles. They came back for physical examinations every 5 years, and obituary notices were screened to determine if any of the subjects had died. This study has continued to run, now following the children and grandchildren of the original subjects [1].

After 5 years, there were enough subjects in the trial who had suffered heart attacks that the statisticians were able to look at potential predictors of a heart attack. They wanted to look at the relationships among lifestyles, heredity, and physical conditions versus the probability of having a heart attack, but they could not use a multilinear regression for a simple reason.

They were attempting to estimate a probability, a number between zero and one. A straightforward multilinear regression could easily produce "predictions" with probabilities greater than 1.0 or less than 0. However, if they looked at the **log-odds**, this would not happen. The log-odds?

$$\text{Log-odds, or logit} = \log_e\left(\frac{p}{[1-p]}\right) \qquad (6.1)$$

where p is the probability.

This is a very compact line of notation. For those who are not used to working with mathematical formulas, let's take it apart. We start with the probability of the event. Since we don't know its value to begin with, we represent it by a letter, p. We don't want to create a regression formula for p that will allow us to conclude that p is less than zero or greater than one, since it is a probability. We can be sure that p will remain less than one if we use the odds instead of p. The **odds** are defined as $p/(1-p)$. The odds go from zero (when $p = 0$) to infinity (when $p = 1$). If we use the odds, $(p/(1-p))$, we don't have to worry that the regression estimate will get too big, but we still have to keep it from going below zero. As a next step, we take the logarithm of the odds:

$$\text{Log}_e\left[\frac{p}{(1-p)}\right]$$

Recall that the logarithm is defined as follows:

$$\text{If } y = e^x, \quad \text{then } x = \log_e(y)$$

This takes care of keeping our function in the right range, because the logarithm of zero is minus infinity. It introduces our friend from the Poisson distribution, Euler's "e." To people, like me, who love mathematics, one of the charms of the subject is that, each time you do something new, you open a door onto a whole new field of research.

The statisticians on the Framingham Study could not just stop with the logit. Their second problem was with the measurement that is made on each of the subjects of the study. It is not like weight or blood urea nitrogen. It is either 1.0 (if the subject had a heart attack) or 0 (if the subject did not have a heart attack). They were trying to determine the probability of a heart attack by looking at the number of patients who had heart attacks versus the number who did not. To avoid getting nonsensical probabilities, they used the following version of the basic statistical model:

$$\text{Logit}(p) = \beta_0 + \beta_1 x_1 + \beta_2 x_2 + \beta_3 x_3 + \ldots \beta_k x_k + \text{error} \quad (6.2)$$

Each of the x values (x_1, x_2, x_3, x_4, ..., x_k) is some number that was observed on each subject at baseline. These might include body mass index, dummy variable if the subject smoked, systolic blood pressure, fasting blood sugar, and so on. Among the baseline variables was a newly developed blood measurement known as total blood cholesterol. The parameters (β_0, β_1, β_2, β_3, ..., β_k) are unobservable parameters that have to be estimated from the observed data.

The probability of an observation (heart attack or no heart attack) can be written in a complicated mess of algebra with all these parameters. We insert the observed values (either zero or one for each subject), and we end up with a likelihood. It would be extremely difficult to find the parameter values that maximized such a likelihood, using only pencil and paper, and a desk calculator. However, in 1954, when the first 5 years of data became available, digital computers had arrived on the scene. The sophisticated programs and massive speed and memory of the modern computer were still in the future, so it took a great deal of effort to adapt the primitive machines of the time to find the maximum likelihood estimates of the parameters.

When this logistic regression was run on the first 5 years of data using only the subjects more than 40 years old, from the Framingham Study, the most important factor in the resulting analysis, the factor that had the greatest influence on whether the subject would have a heart attack, was whether the subject was male or female. The second most important factor was whether the subject's father had had a heart attack. The third most important factor was whether the subject was a smoker. The fourth most important factor was whether the subject was tall and thin or short and squat. The fifth most important factor was whether the patient had uncontrolled high blood pressure. When the study was being organized, someone suggested that they include among the blood chemistries a newly discovered fraction called cholesterol. It turned out that high cholesterol was also a predictor, albeit a weak one.

The Framingham Study introduced logistic regression to the medical community. This method of examining data has proven so useful in medicine that many issues of the major medical journals will have at least one article that uses logistic regression.

One linear model based on a function of the observations, that is widely used in sociology, is the log-linear model. Here, the data are arranged in tables showing the frequencies of some event (e.g., incarceration) in terms of different characteristics, like gender, socioeconomic class, or religion. The log-linear model allows the logarithm of the probability to be written as a multilinear function of the categories in the table. Although log-linear models are often taught as part of a course on dealing with categorical data, they are so important in sociology, that many sociology departments at major universities will have separate courses dealing with the use of log-linear models.

6.1 SUMMARY

Not all problems can be solved with multilinear regressions. When the "predicted" value of the primary measure, y, can go beyond the possible values that y could equal, it is useful to construct a multilinear model for some part of the probability function of the error. An example of this is the use of the logit ($= \log_e(p/[1 - p])$, and the log-linear model.

REFERENCE

1. http://www.ncbi.nlm.nih.gov/pmc/articles/PMC1449227/ (last modified Apr 2005).

CHAPTER 7

Correlation versus Causation

C HAPTER 2 OF THIS book introduced the basic statistical concept:

$$\text{Observation} = \text{truth} + \text{error}$$

The perceptive reader may have realized, at this point, that, with regression, the "truth" has turned into a formula involving other observations taken at the same time and place. Consider the prediction formula for forced vital capacity (FVC) for males:

$$\text{Predicted FVC} = 2.49 + 0.043\,(\text{height}) + 0.029\,(\text{age})$$

We can use this formula to predict the "normal" FVC for a patient, but what does it mean? Is FVC "caused" by height and age? If not, why does the FVC change with height and age?

The Harvard Nurses Study has been following 121,700 female nurses since 1976 and a second cohort of 116,000 since 1989. Upon entering the study, each subject filled out a questionnaire regarding her lifestyle. When the first group reached their 50s, the analysts for the study looked at the characteristics of lifestyle that are associated with high blood pressure. Our friend logistic regression came into play.

One of the baseline factors that was predictive of high blood pressure was whether the nurse made regular use of sunscreen. Nurses who reported regular use of sunscreen had fewer incidents of high blood pressure. Sunscreen? Sunscreen is an ointment that is not absorbed through the skin and that blocks the passage of ultraviolet light onto the cells of the skin. Does sunscreen prevent high blood pressure? If so, what is the pharmacological mechanism?

The explanation the directors of the study settled on was that use of sunscreen and occurrence of high blood pressure later in life are correlated, not because a person can prevent high blood pressure by using sunscreen but because nurses who regularly use sunscreen are more careful with their health and are also more likely to follow reduced salt diets and engage in regular exercise. Both of these lifestyle practices can help prevent high blood pressure.

Spurious correlations like this can be found in the medical, sociological, psychological, and epidemiological literature. During 1900–1914, the yearly incidence of divorce and the registration of motor cars in Great Britain were highly correlated. But, divorces did not cause people to go out and buy motor cars, and the purchase of a motor car did not cause couples to divorce. Both were functions of increasing wealth and urbanization. The same explanation holds for the high yearly correlation in post-World War II Wales of young male suicides and automobile registrations. Wales underwent a rapid urbanization in the years following World War II, and it has been well established (among immigrants to the United States and newly developed cities in Africa during the twentieth century, for example) that urbanization leads to an increase in suicides among young males.

When evidence began to mount suggesting that cigarette smoking "causes" lung cancer, the Tobacco Institute (a "research" organization funded by the cigarette companies in 1958 and dissolved in 1988) ran full-page advertisements in newspapers. The advertisements pointed out that this relationship did not prove that cigarettes cause cancer since it was only a statistical correlation.

Setting aside the questions involved in cause and effect, let's examine the nature of statistical correlations.

Table 7.1 and Figure 7.1 display the relationship between heights and weights of 13 women living on Navaho reservations in the United States in 1962. The two measures are clearly related. If we knew nothing about the weights, our "best" estimate of the mean height for women like these would be the average of the 13 women or 141.2 cm. There is considerable scatter about that average. The shortest woman was only 114 cm while the tallest was 157 cm.

The usual statistical measure of that scatter, the sample variance, equals 242.4 cm^2. But, knowing a woman's weight provides some information about her height. The correlation between

TABLE 7.1   Relationships between Heights and Weights of Navaho Women in a Sociological Study Published in 1962

Weight (kg)	Height (cm)
19.0	113.6
21.2	119.0
23.1	123.9
25.6	129.9
28.3	134.9
31.3	140.2
34.6	143.9
39.0	150.3
43.6	153.1
47.7	155.3
51.1	156.3
52.3	157.4
54.0	157.5
Average: 36.2	**Average: 141.2**

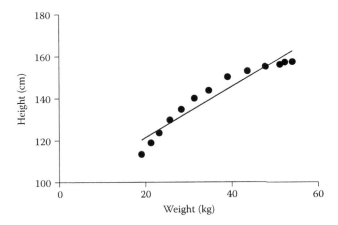

FIGURE 7.1   Heights and weights of 13 Navaho women measured in 1962.

height and weight for these data is 0.963. This alone does not tell us much, but, if we fit a linear regression where y is the height and x is the weight (see Figure 7.1), then the variance about that line is much smaller. In fact, it is reduced by the square of the correlation (usually referred to as R^2). Using a linear regression, we can "explain" 92.7% of the variance.

This idea of being able to reduce the variance of y by knowing the values of the x variable used in the regression carries over to multilinear regression. If you run a multilinear regression using standard commercially available software, it will present you with the value of R^2 for the entire regression model. Different fields of science have different standards for what is an acceptable value of R^2. In chemistry or physics, the proposed regression is considered useful only if it can account for 90% or more of the variance. In biology or medicine, you can publish an article in a reputable journal if you can account for 30%–40% of the variance. In sociology and psychology, subjects that deal with human behavior, I have seen papers where 20% was acceptable.

Note that when we reduce the variance of the y variable, we are not saying that the multilinear regression model shows us what "causes" y. We are saying that, by using the regression model, we can predict the value of y with less uncertainty.

The use of R^2 to describe the reduction in variance can be extended to individual elements of the regression. Mathematically, if we hold all but one of these baseline variables at their average, we can calculate the percentage reduction in variance associated with that one variable (e.g., weekly amount of exercise). Thus, while we are still not dealing with "cause," we can use the calculations of correlation to tell us how useful knowledge of portions of the regression is in predicting values of y.

The calculations of R^2 can be made for multilinear regressions but not when we are dealing with a modified y variable, like the logit. Other measures (such as the percentage reduction in probability) have been proposed to estimate how well the predicted values fit the observed values of y. Some are available in commercial statistical software.

But, what about "cause?" We know that we can better predict the occurrence of lung cancer if we have knowledge of the subjects' smoking habits. But, does this statistical correlation prove that smoking "causes" cancer?

The problem with this question is that statistical correlation is well defined. We can calculate R^2 (or its equivalent in more complicated models) because there is a well-defined mathematical computation that provides us with a number.

But, "cause and effect" is not well defined.

That is a foolish statement; I have had students tell me. Everybody knows what "cause and effect" means. Alright, I tell them, then please define it for me in terms of more primitive concepts. Can you define "cause and effect" without using a synonym for cause? It cannot be done. The Scottish historian and philosopher David Hume (Figure 7.2) discovered this in the eighteenth century. The Greek philosopher, Aristotle, had proclaimed that you do not understand something unless you know its cause. Over 2,000 years later, Hume asked, what does that mean?

FIGURE 7.2 David Hume (1711–1776), a Scottish historian and philosopher who discovered that "cause and effect" has no clear meaning.
(Courtesy of Shutterstock.com.)

Hume concluded that we cannot define "cause and effect" as we
define something specific like iron. However, vague as the idea of
causation is, it is often useful in our day-to-day life. Hume concluded that whether we believe that A "causes" B depends upon
our finding cases where A preceded B in time. If we find that people who have lung cancer were smokers, we can conclude, but only
tentatively, that smoking causes lung cancer. Every example of a
smoker who has lung cancer reinforces our conclusion. And, every
instance where we find a smoker who did not get lung cancer, or

where we find someone with lung cancer who was not a smoker, reduces our belief that smoking causes lung cancer.

Hume's "definition" is not very satisfying. To Hume, "cause and effect" is nothing more than a tentative suggestion. This is a rather squishy "definition" compared to the way we define correlation. But, it was the best Hume could do.

There are four other definitions of cause and effect that I know of.

One is the primitive definition. "Cause and effect" occurs when a willful force engages in some activity with the intent to produce an effect. The caveman hurls his spear at a wooly mammoth with the intention of killing it, thereby causing its death. The local factory has closed because a conspiracy of bankers has desired to destroy the company. The winds blow because the god of the winds has released them from his cave with the intent of blowing down trees. My brother, out in his flimsy boat, has been struck by lightning because the god of the water (or the one true God) has desired to do so. Most conspiracy theories are based on this primitive concept. Terrible things happen to the nation or the economy because a conspiracy of politicians(?), bankers(?), foreign merchants(?) have caused it to happen. But, there is no evil willful force causing cigarette smokers to have lung cancer. The nicotine or impurities in the smoke may lead pharmacologically to lung damage that may develop later into cancer, but the impurities in the smoke do not desire the smoker to have cancer. Thus, this definition is not useful in the case of smoking and cancer.

The second definition is due to the nineteenth century German bacteriologist, Robert Koch (1843–1910). When scientists like Koch began to suggest that many illnesses are caused by infectious bacteria, Koch proposed a set of postulates that could be used to prove that a specific species of bacteria caused the disease.

Koch's postulates are that an agent X causes disease Y if and only if

1. Whenever agent X can be cultivated from the blood or tissue, the person has disease Y.

2. The agent X can be cultivated from the blood or tissue whenever a patient has disease Y.

3. The agent X, having been cultivated this way, can be injected into an animal, and the animal then gets disease Y.

4. The agent X can be cultivated from the blood or tissue of that animal.

5. The agent can then be injected into another animal, and that second animal gets disease Y.

Koch's postulates are useful for identifying infectious bacteria or viruses but are useless if we want to determine whether cigarette smoking causes lung cancer or whether failure to use sunscreen causes high blood pressure.

A third definition was given by R.A. Fisher (Figure 7.3), the great genius of the first half of the twentieth century who laid the mathematical foundations of modern statistical theory.

In Fisher's definition, cause and effect can be determined only within the framework of planned experiments. If we want to determine whether some treatment (call it A) produces some effect (call it B) on plants, animals, or people, we assemble a collection of experimental units and assign either treatment A or not treatment A at random to these units. We observe the results (usually in terms of a number for each unit) and calculate the average difference between the units given A and those not given A. Since we assigned units at random, any other random assignment was equally probable. So, taking the results of our experiment, we pretend to reassign treatments at random, and compute the difference in average outcome. We do this for all possible random assignments and calculate the average difference for each one. This gives us a probability distribution of average differences if treatment A had no effect. If the difference for the assigned treatments is among the least probable for all possible assignments, then we declare that treatment A caused a difference. Fisher called this complicated approach "inductive

FIGURE 7.3 Sir Ronald Aylmer Fisher (1890–1962) the genius who laid much of the groundwork of modern mathematical statistics. (Public domain.)

reasoning." Note that Fisher's definition can be used only in the framework of a randomized experiment. So, we could not use Fisher's definition in deciding whether smoking causes lung cancer.

During the early years of the twentieth century, a group of mathematicians (primarily in Italy, Germany, and England) put the logic of mathematics on a firm setting where all assumptions were clearly defined, and all conclusions followed from application of only three logical concepts, "and," "or," and "not." This is called symbolic logic. The closest they could come to "cause and effect" is called material implication. An event A implies an event B if it is not possible for B to occur when A has not occurred. Thus, if we can find a nonsmoker (not A) who has lung cancer (B), smoking does not cause cancer in the sense of material implication.

A fifth "definition" is due to Bertrand Russell (1872–1970) who was one of the mathematicians involved in the development of symbolic logic. Russell called "cause and effect" a "silly superstition."

There is another definition of "cause" that involves probabilistic reasoning. The major proponent of what is called "causal analysis" is Judah Pearl (b. 1936) at the University of California, Los Angeles. Causal analysis looks at two events, A occurring before B. If the existence of A increases the probability that B will occur, then this increase in probability is examined in causal analysis. Causal analysis uses the mathematical theory of graphs, which takes any explanation of the idea far beyond the scope of this book (see Reference [1]).

The next time someone tells you that some relationship is only a statistical correlation, and that it has not been proven to be a cause, ask that person "please define 'cause.'" Then sit back and watch the fireworks.

7.1 SUMMARY

The degree to which one variable can be predicted from another can be calculated as the correlation between them. The square of the correlation (R^2) is the proportion of the variance of one that can be "explained" by knowledge of the other.

The concept of $100 \times R^2$ as the percentage of variance accounted for extends to multilinear regressions, so we can calculate how well a particular model fits the data. We can also calculate R^2 for a specific element of the multilinear regression and determine how well knowledge of that element predicts the outcome variable. Although one cannot calculate R^2 directly for modified linear models like the logit, similar measures have been proposed for each of them.

Correlation is not equivalent to cause for one major reason. Correlation is well defined in terms of a mathematical formula. Cause is not well defined. In fact, David Hume showed that cause and effect cannot be defined in general. There are five possible restricted definitions of "cause and effect":

1. The primitive definition that a willful force did something with the intent of producing an effect.

2. Koch's definition that holds only for bacterial or viral infections.

3. Fisher's definition that holds only in the framework of randomized controlled experiment.

4. Material implication from symbolic logic.

5. Russell's "definition" that "cause and effect" is a silly superstition.

REFERENCE

1. A description of causal analysis, http://ftp.cs.ucla.edu/pub/stat_ser/r350.pdf (last modified Sep 2009).

Regression and Big Data

A S OF THIS WRITING, one of the most widely discussed topics in the statistical literature is how to deal with big data—the huge amount of information that is being accumulated on Internet servers and the potential value of fishing in that data for unexpected relationships. Let's see how this problem fits into the general statistical model:

$$\text{Observation} = \text{model} + \text{error}$$

Early in my career at Pfizer Central Research, we were developing a new type of drug for the treatment of high blood pressure. The first really effective drugs for high blood pressure had been developed in the 1950s and worked by interfering with the hormones that the kidney produces to control the force of the heart. These were called diuretics. They were effective for about 25% of patients. The drug we were investigating was an alpha-adrenergic blocker, which affected the muscle walls of arteries and was expected to prove effective in patients whose blood pressure did not respond to diuretics.

But, this was during the 1960s, soon after amendments to the food and drug laws of the United States had given the U.S. Food and Drug Administration (FDA) the tools it needed to require that drug companies produce more research before they could market a new drug. As a result, our new drug candidate had to be subjected to additional amounts of toxicological studies involving more than one species of animal. The dog studies showed that our new drug was associated with testicular atrophy at high doses.

This might have been something unique to dog metabolism, but who could be sure it would not hold in humans? Some of the reviewers at the FDA proposed that we take frequent measurements of testosterone levels in the blood of all males exposed to the drug. However, at that time, the assay for testosterone was not well standardized, and the company would have to first engage in a large and expensive study to establish standard methods of assay. As a compromise, the company and the regulators agreed that it would be sufficient to collect data on urine levels of 17-ketosteroids, a metabolic breakdown product of testosterone, which had a recognized standard assay already.

And so, the human studies began. They ran for several years, as they extended from initial (Phase I) studies in healthy volunteers, to (Phase II) studies aimed at developing appropriate dosing patterns to (Phase III) studies that established the efficacy and safety of the new drug. No cases of testicular atrophy showed up in any of these studies.

In the meantime, the determinations of 17-ketosteroids were accumulating. When the data from all the studies were being tabulated and analyzed, we had to deal with these measurements. We first tried to compare the values we had with the normal range. Recall from Chapter 4 that the normal range is the range of values that cover 95% of the determinations found in normal healthy males.

Only, there were no normal ranges. These breakdown products of testosterone had been examined for their chemical structure and for their reactions with other hormones both in vitro and in vivo, but no one had taken a large number of healthy men and determined the normal range.

The company librarians searched through the scientific literature for some example, somewhere, where the 17-ketosteroids had been measured on many healthy men. They found a paper in an obscure small circulation journal entitled, *20,000 Determinations of 17-Ketosteroids*. With much effort, we obtained a copy of this paper. (This was before the Internet and the ease we now have of finding most scientific papers.) The author of the paper had measured the 17-ketosteroids in his own urine three times a day for 20 years.

When running a multilinear regression, there are two aspects of the data usually called N and p. N is the number of statistically independent individuals on whom the measurements are made, and p is the number of elements in each of these independent observations. It is sometimes called the dimension of the vector of observations. In running a multilinear regression, it is necessary that N be much greater than p before we can assign any degree of precision to our estimates. If the author of this paper had taken 17-ketosteroid measurements in the urine of 20,000 men, we would have been able to establish the normal range with great precision. Instead of $N = 20,000$ and $p = 1$, this paper had $N = 1$ and $p = 20,000$.

This is an extreme example. However, many of the files of Big Data suffer from a "big p little N" problem. One person sitting at a laptop computer will strike the keys of the keyboard, and, somewhere, another computer is storing each of those strokes and a great deal of information associated with each stroke. If the same person comes back at another time, and the strokes and associated information are recorded, has N been increased by 1? Or has N remained the same and p been doubled? And, is the information needed to distinguish the difference in the file?

Even if great care is taken to identify the statistically independent units being stored and the material being stored is carefully pigeon-holed into predetermined categories, Big Data will suffer from the "p a little too big and N not big enough" problem.

But, something should be done with Big Data. In the Framingham Study, the initial subjects were followed for years, measured every 5 years, note taken of work habits, lifestyle, food eaten, diseases encountered, socioeconomic variables, and many more. Even while using hundreds of thousands of women, the Harvard Nurses Study keeps accumulating more and more information about each one. If we want to use Google searches to identify disease patterns in a community, there will be hundreds of thousands of web sites that might have been searched. When accumulating data, p has a way of mounting, but we cannot help but think that, somewhere in that mess of information is something that might be useful.

With the computer to aid us, a number of statistical procedures have been proposed for dealing with the "big p little N." Some methods look at each of the elements of the p "explanatory" variables, examine its predictive value, and keep only the most predictive ones. Other methods apply a penalty function to the algorithm, so the "best" collection is kept small by penalizing the definition of "best" if it produces too many explanatory variables. Other methods look at the interrelationships among the p explanatory variables and find a small set of combinations that "describe" the set of p values "best." Before the Internet began to create large data files, the medical profession was running studies where a relatively small number of patients were examined and measured in many different ways multiple times during the study. John Tukey had some experience dealing with these data files. At a statistics meeting, I heard him say that there were really no more than five dimensions to most medical data—by this, he meant that all the information could be gained from looking at five carefully determined functions of an entire set of observations on a given patient.

The "big p little N" problem is currently generating a great deal of theoretical research. Eventually, as has happened with every previous area of research, effective mathematical tools will emerge and become part of the range of "ordinary" statistical methods.

There is still another problem with big data that results from the logic of the mathematics. Theoretically, it might be possible to find a set of variables for which the x and y values have a perfect fit, and for which $R^2 = 1.00$. The fact that we have a finite number of observations (N is not infinite) means that, as you increase the number of variables in the model, the value of R^2 will begin to increase only because you are approaching this limit where the model is so complicated that everything we have observed "fits" perfectly. Another set of data drawn from the same distribution for the error will produce another but different perfect "fit." Neither perfect "fit" provides useful predictions for future data. To avoid being led astray by spurious perfect fits, we reduce R^2 through the use of a "penalty function" governed by the number of elements in the regression. With this penalty function thrown in, you get what is known as the "adjusted R^2." It is the adjusted R^2 that is reported in most commercial statistical software, usually as "adj. R^2."

All methods of dealing with big data require a vast number of mind-numbing, tedious, boring mathematical steps. Fortunately, we have this big unimaginative servant, the computer, that does not seem to mind doing these things over, and over, and over, and so on. As the twenty-first century advances and new bright young statisticians exploit the subtleties of the modern computer, I expect that the "big p little N" problem will be dealt with in ways yet to be explored.

Once we have reduced the problem from "big p little N" to "a moderate sized N and p a little too big but not yet bigger than N," we might look among these variables and say, "Aha, smoking seems to be associated with pancreatic cancer, as does caffeine intake from colas (but not from coffee), as does socioeconomic level, as does whether the patient's right big toe is bigger than his/her left big toe, as does the phase of the moon on the patient's 25th birthday."

The reason why you begin to see spurious correlations is due to a problem first examined by the Italian economist Carlo Bonferroni (1892–1960) in 1936. If you look at a lot of data, you are bound to find some peculiarities of those data that are unique to the set you have on hand and have no predictive value. Francis Anscombe (1918–2001), who founded the Statistics Department at Yale University, called these weird little coincidences "will of the wisps."

So, how does one avoid pulling out one of Anscombe's will of the wisps? Bonferroni's solution was to make the criteria for choosing a variable very stringent and make it more stringent the more variables you consider. His formulas for invoking that stringency are known as "Bonferroni Bounds." Bonferroni's criteria are so stringent that, if you have more than 20 variables to consider it is almost impossible for any one variable or a group of them to be chosen.

An example of what can happen when you fail to heed Bonferroni and Anscombe's warning is *The Bible Code*, by Drosnin [1]. The author has gone through the Hebrew letters of the Five Books of Moses and combined them by skipping every third letter, then every fourth letter, and so on, looking at one different combination after another. Eventually, he found the name of one of the great kings of the Kingdom of Judah. Then, he found warnings of events that occurred in the nineteenth and twentieth centuries of the current era.

In his hands, the Hebrew letters of the Pentateuch seems to predict many events that happened thousands of years after Moses (but always before the date that Drosnin found them, none predicting future events).

So, how does one avoid falling for one of Anscombe's will o' the wisps? One tried and true technique has been available, since it was first proposed by R.A. Fisher in the 1920s. It has been proposed again and again by leading statisticians like Jerome Cornfield (1912–1979) and John Tukey.

Before you begin to examine a mass of data, you pull out a small randomly chosen subsample. Tukey suggested 10% of the data. Fisher suggested that the sample be just big enough to allow for a regression analysis. The Harvard Nurses Study identified a cohort of about 1000 subjects as their random subset. You, then, do all your exploratory analysis on this small subset. You run stepwise regressions. You ignore Bonferroni's stringent criteria. You look through the data and identify subjectively interesting relationships. Tukey called this "exploratory data analysis." You use it to propose an interesting set of possible relationships. You don't let the output of a statistical program guide you completely, but you use your knowledge about the situation being examined to pick some hypotheses generated by this exploratory analysis and to reject others.

Eventually, you have a small number of reasonable, sensible models that are suggested by this randomly chosen exploratory sample. You, then, go back to the rest of the data (the remaining 90% in Tukey's version) and run formal statistical analyses to identify which of these models holds.

There are other problems with Big Data. In any large data set, there are bound to be inconsistencies, misclassifications, missing data—in other words, errors, blunders, and possibly lies. These problems with individual items occur in any data set, but they are often hidden in a large mass of numbers even when these numbers are generated out of computer interactions—but the cleaning of data is a topic for another book.

8.1 SUMMARY

There are two dimensions to a large data set, N, which equals the number of statistically independent units generating that data, and p, which equals the number of items generated by each independent unit. Traditional statistical methods require that N be much greater than p. In our modern era of big data generated by use of the Internet, it often happens that p is sometimes almost as big as N or even greater.

Since it is necessary to reduce the size of p in order to run useful multilinear regressions, methods have been developed to identify the elements of each independent unit that are useful in predicting observations of variable, y.

Because it is theoretically possible to "fit" the data exactly with enough elements in the regression, the criteria for leaving elements in or out is the adjusted R^2, where the calculated R^2 is adjusted downward for increasing numbers of elements in the regression.

Within any large data set are what Anscombe called "will o' the wisps," apparently strong relationships that are unique to that data set and have no predictive value. Bonferroni bounds are often used to reduce the chance of finding one.

A better procedure for avoiding will o' the wisps and producing spurious conclusions is to select a small subset of the data at random and use that subset for exploratory analysis. The rest of the data are left for formal statistical analyses to identify those elements that have predictive value.

8.2 FOR THOSE WHO WANT TO SEE THE MATHEMATICS

Bonferroni's bounds are derived as follows.

Hypothesis testing is the definitive method used to identify results that are better than would happen by chance. We pick a small probability, called α and usually set it at 0.05. We, then, examine the output of an experiment. Even if there is no difference in the effect of treatment, we might have observed an output favorable to the hypothesis that treatment has an effect. But, under hypothesis testing, only if the probability of such an observation is less than α, do we declare that the difference is real.

Bonferroni asked what happens if you make more than one such decision. Suppose you make 10 decisions, each one based on a 5% probability associated with mere chance. For each of these 10 decisions, the probability that we will not detect an "effect" when there is no effect is

$$1 - .05 = 0.95$$

If all the decisions are statistically independent, then the probability that all 10 will be found to have no effect is

$$(1 - .05)^{10}$$

Thus, if we examine 10 different effects where none of them is real, the probability that at least one will be flagged is

$$1-(1-.05)^{10} = 1-[1-(10\{.05\}+45\{.05\}^2-120\{.05\}^3\dots\{.05\}^{10})]$$
$$= 10(.05)+(\text{a lot of small positive numbers})$$

If we ignore all those small positive numbers, the probability of at least one false positive is less than or equal to .50. It could be that several of the tests are not independent and might be asking the same or similar questions. Then, the probability of no false positives may be less than $(1 - .05)^{10}$. We have to take this as an upper bound. But, if two tests are asking the same question, we really should not have been using both.

In general, if you make K decisions, each at the error level of α, then the probability of making at least one false decision is less than or equal to

$$K\alpha$$

And so, if you make K decisions, you can declare an effect only if the probability of having seen it by chance is less than

$$\alpha/K$$

This seems like an obvious and simple finding. Anyone with 2 years of high school algebra knowledge could prove it. However, in 1936, mathematical statistics was just beginning to become a subject of interest. Bonferroni was the first person to publish this obvious result. *Moral:* If you want to become famous, find an emerging field of academic interest and quickly publish something that is obvious before anyone else does.

REFERENCE

1. Drosnin, M. (1998) *The Bible Code*, New York, NY: Touchstone Press.

Blunders

CHAPTER 9

Contaminated Distributions

D URING WORLD WAR II (and before radar was developed
sufficiently to provide accurate measurements of distance),
the U.S. Navy used an optical rangefinder on its ships. It consisted
of a bar about 10 feet long (longer or shorter depending on the size
of the ship) with openings at the ends and mirrors that allowed
someone peering through a lens in the center to see two images of
the target whose distance is sought.

The initial idea was to rotate one of the end mirrors until the
observer saw both images merge. Using simple geometry, the
angle of the movable end mirror and the length of the range finder
bar could be used to compute the distance to the target.

Unfortunately, the bar was short (measured in feet) relative to
the distance to the target (measured in miles), and very slight slippage in the gears running the rotating mirror could lead to large
errors in the range. During World War I, the German Navy developed a more accurate model that changed the view presented to
the person running the rangefinder to a three-dimensional image
of the target (as viewed by "eyes" at either end of the bar). A set of

black diamonds was presented, placed over the three-dimensional image of the target. The largest one was in the center and the others were seen before or behind the target. The idea was to turn the dial, adjusting the angle at one end of the bar until the large diamond could be seen over the target.

This more sophisticated version was adopted by the U.S. Navy at the beginning of World War II. Figure 9.1, taken from the U.S. Naval Gunnery Manual of the period [1], shows images presented to each eye. Figure 9.2 shows a breakdown schematic of a rangefinder.

Statistical models were being widely used at this time, so it was felt necessary to establish the probability distribution of the error involved in using such a device. At a Navy yard in the United States, a new rangefinder was mounted on a tower, and a ship was anchored at a known distance from that tower. A large number of sailors were brought to the site. One by one, they climbed the tower and used the rangefinder to determine the distance to the target ship.

A petty officer was put in charge and given a table of random numbers. After each sailor used the rangefinder to measure the distance, the petty officer would record it, and, using the next

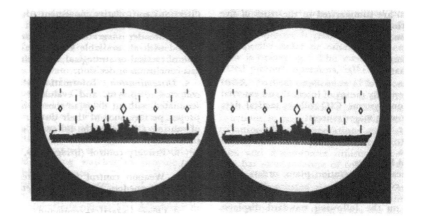

FIGURE 9.1 Stereoscopic view of a target as seen through an optical rangefinder. To see it in three dimensions, place a cardboard edge between the two views and let each eye see a different view.

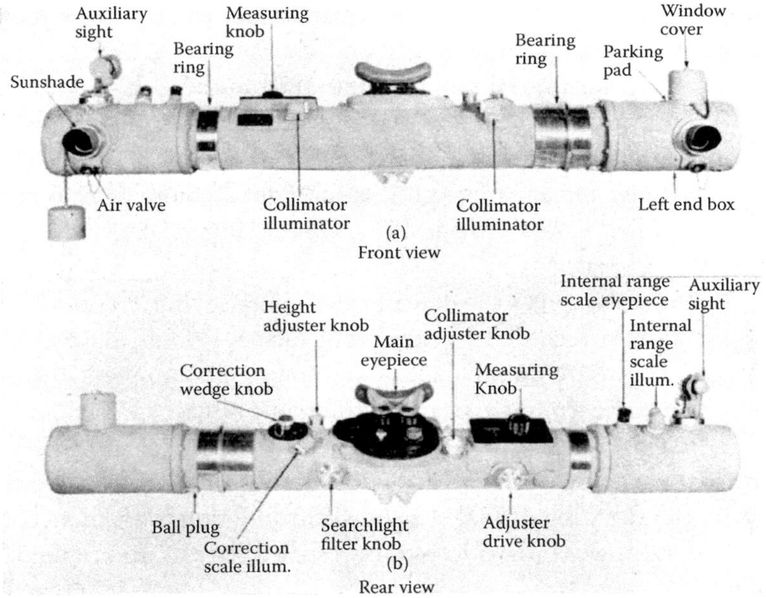

FIGURE 9.2   Mark 58 optical rangefinder used by the U.S. Navy prior to the development of radar.

number on the table of random numbers, he would twist the dial, setting it at that random entry. This way, each sailor's ranging would not be influenced by the previous sailor's value.

When the data were analyzed, they had a peculiar pattern. Most of the ranges were bunched up around the true value of the distance or only slightly off. However, there were some range values scattered all over the possible values, almost as if they had come from a table of random numbers and had nothing to do with the distance being measured.

It turns out that around 5% of people have amblyopia or "lazy eye." Although they can see perfectly well out of both eyes, their brains suppress the image from one of their eyes, so they cannot see three-dimensional presentations. Imagine a sailor with lazy eye climbing the steps up to the rangefinder and being told by the petty officer to peer into the device and place the big black diamond over the ship being ranged on.

"Over?" the sailor asks looking into the rangefinder.

"Yes," says the petty officer. "Don't you know what 'over' means?"

"But," the sailor begins, "I don't see…"

"Don't give me any more guff, sailor. I got 25 more guys to put through this before lunch. PUT…THE…BIG…DIAMOND…OVER THE SHIP!"

So, the sailor twists the dial and steps back saying, "It's done."

The ranges recorded by the petty officer belong to what is known as a "contaminated distribution." The numbers being observed come from two or more different sources.

(I first heard the story of the optical rangefinder at a meeting of the American Statistical Association. I have been unable to verify the story, but it makes a good introduction to the concept of a contaminated distribution. I present it here as an academic legend.)

Table 9.1 displays 60 random numbers generated from a standard normal distribution with 10% contamination from a uniform distribution where all the numbers between −10 and +10 are equally probable. This is the sort of pattern that would have shown up in the rangefinder trial, except that, with the rangefinder, there was only 5% contamination. If I had generated numbers with only 5% contamination, it would have been difficult to see the effect. Even with 10% contamination, only the largest four values, 9.80, 8.30, 5.90, and 5.40, and the smallest two values, −9.20 and −9.10, seem out of place. In Chapter 10, we will examine this type of contamination in greater detail.

The effects of contamination can be very subtle. Before the advent of automatic blood pressure reading machines, the person taking a patient's blood pressure was trained to listen carefully to the different sounds in the sphygmomanometer (yes, that is what that set of tubes and bladder is called). The cuff is pumped up to block the blood flowing through the artery over which the bladder is placed. Then, the pressure is released slowly. When the observer

TABLE 9.1   Example of Data Where There Is a 90% Chance the Data Came from a Normal Distribution with Small Variance and a 10% Chance That They Come from a Probability Distributed Evenly from −10 to +10

−9.20	−1.40	−0.48	0.16	0.63	1.39
−9.10	−1.38	−0.31	0.16	0.66	1.49
−3.90	−1.35	−0.29	0.18	0.77	1.68
−2.70	−1.34	−0.15	0.20	0.91	1.90
−2.40	−1.33	−0.15	0.23	1.03	2.27
−1.81	−1.19	−0.11	0.29	1.04	3.50
−1.79	−1.13	−0.01	0.38	1.18	5.90
−1.51	−1.01	0.00	0.38	1.25	6.40
−1.50	−0.85	0.06	0.42	1.37	8.30
−1.42	−0.69	0.14	0.46	1.39	9.80

first hears a "tink, tink, tink," that is the beginning of the blood flow and marks the pressure produced by the heart in systole. The observer continues to listen until there is a rushing sound. That is the flow of blood against the back-pressure of the cardio-vascular system and defines the heart in diastole.

However, not everyone does this the same way. A nurse measuring blood pressure will usually take the numbers as they appear. Often, a doctor measuring blood pressure will have pathological problems in mind that might occur and will sometimes question the diastolic value if he or she thinks it does not "agree" with the systolic value. In such a case, the doctor pumps it up again to get the "right" answer. As a result, blood pressures measured by doctors tend to have more highly correlated systolic and diastolic values than blood pressures measured by nurses.

In such a case, the contamination does not show up as data with a different mean or variance but as data with different correlations.

What can be done with a contaminated distribution? Is there any way to separate the contaminating data from the "real" data? The answer depends upon the nature of the contamination.

Sometimes the blunders, the contamination, seem to be "obvious," but it may not be true. I was once involved in analyzing the weights of rats used in toxicological experiments. A mature rat of the strain we were using can weigh between 200 and 300 grams, with the females slightly less than the males. I had a file of the raw data and was streaming through it, looking for obvious anomalies. When you analyze a lot of data, you learn (often the hard way) that even the most carefully developed set of data will have problems that were not anticipated. One way to protect yourself from embarrassing surprises is to look at the actual data before running regressions or finding averages or sample variances.

And, sure enough, there it was among the mature weights of the rats, a female rat weighing 2000 grams. It was obviously a mistake, someone had slipped a decimal point, or the automatic scale got stuck. Something happened on the way from that rat to my data file. I went to the toxicology lab, which had generated the data, to see what had happened.

Oh, they all remembered her! She was so fat that she propelled herself on her belly and her sides stuck out between the bars of her cage. Was this a blunder? Did this fat female rat represent something from a contaminating distribution? It depends upon what we were trying to find. If we were looking for the average of all the rats, this one outlier might skew our estimate too high. But, if we were looking for the variance, then it was important to know that, on rare occasions, we will have a 2000-grams female rat.

9.1 SUMMARY

Contaminated distributions occur when some of the data are generated by the situation being examined, but some of the data come from an irrelevant situation. Contaminating distributions can increase the variance, change the mean, or even effect the correlation. It is difficult to determine whether a single outlier has been generated by a contaminating distribution.

REFERENCE

1. U.S. Department of the Navy, Bureau of Ordnance (17 Apr 1944) *Optical Equipment—Rangefinder, Mark 58, 58 Mod1, 64, 64 Mod 1,* Washington, DC: U.S. Department of the Navy, Bureau of Ordnance.

CHAPTER 10

The Princeton Robustness Study

D URING THE ACADEMIC YEAR 1970–1971, the faculty and graduate students of the Princeton University Statistics Department began one of the most extensive and massive computer-generated **Monte Carlo studies** ever run. Under the leadership of John Tukey (Figure 10.1), they looked at the effects contaminated distributions have on statistical procedures and what adjustments had to be made in standard procedures to protect against these effects. The Princeton Robustness Study was a landmark look at the problem of blunders. Reference [1] is a detailed description of the mathematical methods that resulted from the study. Reference [2] is a lecture by Stephen Stigler aimed at a more general reader [3]. Reference [4] is a retrospective look at the influence this study has had on statistical practice.

What is a Monte Carlo study? This is a study where the computer is used to generate random numbers, filter them through a well-defined probability distribution, and apply different statistical procedures again and again and again...

FIGURE 10.1   John Tukey (1915–2000). A dominating figure in statistical developments of the last half of the twentieth century. (Courtesy of Orren Jack Turner.)

What does the word **robustness** mean in the Princeton Robustness Study? The term robustness had been coined by George Box (1919–2013) at the University of Wisconsin. Box was concerned with how well standard statistical procedures worked when the assumptions behind the mathematical model were not quite correct. For instance, what happens in multi-linear regression if the error is not normally distributed but comes from the class of stable symmetric distributions that Eugene Fama found to fit stock market prices? Box called a procedure that still produced correct answers—a **robust** procedure. He resisted a more specific definition, because he did not want his idea to get bogged down in complicated mathematical sophistication.

John Tukey, at Princeton, thought differently. He wanted to define robust in terms of specific departures from the assumptions.

He wanted to be able to measure how robust different procedures are, so he could compare several and find the "best." To this end, Tukey and his colleagues tried to model contaminated distributions. They dealt with various probability distributions of the blunders.

They quickly realized that the problem was much too general to be addressed, even with the aid of computers. Blunders could come from anywhere, and the contamination of the distribution could twist conclusions in many different ways.

The error distribution is assumed to have a zero mean, be symmetric, and have a small variance. The distribution of blunders could pull the observed values in one direction and destroy the symmetry, or they could be measuring something else and shift the mean far away from what we are trying to measure. They could be symmetric and average zero but be wild shots (e.g., some of the data were recorded with the decimal in the wrong place or the patient weights were recorded as pounds instead of kilograms). They decided to limit their investigation to contaminating distributions that were symmetric and had almost the same mean as what we think we are measuring. The wild values produced by sailors with "lazy eye" fit this model.

They ran a very large number of Monte Carlo studies, where the contaminating distribution, the blunders, were very rare, and then less rare, and then more common, and so on.

Before we consider the results of the Princeton Robustness Study, here is an example from real-life sports that was described by Alexander Meister [5] and supplemented by a private communication from Joseph Hilbe [6]. In track and field competitions, it has always been difficult to determine the exact time of something like the 100-meter dash. Before the introduction of electronic timing devices, the winning time was usually measured by three timers with stopwatches. Since, in sports as in everything else

$$Observation = truth + error$$

the three different timers usually came up with three different times. To reduce the variance of the error, the official time was taken as the median of the three numbers. In longer runs, with only two official timers, the official time was taken as the greater of the two.

Taking the median of the three times produces a measurement where the error is still symmetric but has a smaller variance. It was assumed in these trials that the reduction in variance was sufficient to be reasonably sure of the time within plus or minus one tenth of a second. (The procedure that used the greater of two introduced a bias into the analysis by accepting an error that made the time greater.) Meister points out that, with the introduction of electronic timing the variance of the error was greatly reduced, so we can be reasonably sure that the time was within one thousandth of a second. This raises the question, according to Meister, of whether some modern runner might be "cheated" out of the record books because a previous runner's time had been measured with an error of one tenth of a second or more. One can think of the winner's measured time as having two distributions of error. Before electronic timing, the error variance was larger than after electronic timing. This is the type of situation investigated by the Princeton Robustness Study.

One discovery of the Princeton Robustness Study was that the average of a set of numbers is a very poor estimate of the center of the distribution (the underlying mean) when contamination is there. This finding extended beyond averages. The usual algorithms for estimating a multilinear regression also failed to be robust. To see why this is so, consider the problem of trying to determine the average age of the people in a nursing home. One elderly illiterate man does not know when he was born, but he "figures" that he is at least 110 years old. Most of the others are in their 70s or 80s with a few in their 90s. If you include 110 among the numbers being averaged, you could easily have an average of 93 or 94, when most of the patients are less than 90. Or, consider trying to determine the average salary of the graduates of a particular college 10 years after graduation, when one former student entered a monastery and took a vow of poverty.

Faced with possible contamination by blunders, it is hard to determine whether a particular value is part of the contaminating distribution or a "bona fide" representative of the thing you are trying to measure. Just because a value is an outlier, it doesn't mean it should not be used. Throwing out data that appear to be wild shots can lead to erroneous conclusions.

There is a competitor to the average, as a way of estimating the center of a distribution. This is the median, a number that has half the observations greater than or equal to it and the other half less than or equal to it. Absent contamination, the variance of the average is less than the variance of the median, but, if I had used the median in my analysis of weights of rats, then the 2000 grams female rat would not have pulled my estimate of the mean in its direction.

The median is an example of a more general idea, called the **trimmed average**. For the trimmed average, you order all the data and knock out the five smallest and the five largest. Or, you might knock out the 10 smallest and the 10 largest. It can be seen that the median is an extreme version of the trimmed average.

This is typical of mathematical reasoning. You look at a procedure and try to find a more general way to describe it, so the initial procedure is just one of a large number of related procedures. Then, you try to determine which member of this more general class is the best.

So, what trimmed average is the best? We need more knowledge to answer that. Running their Monte Carlo studies, the Princeton group defined what they called the "break-down" point. If we know the distribution of the error term (probably normal) and the distribution of the contaminating blunders, then we can find a low value (call it A) and a high value (call it B), such that there is a reasonable probability (say >20%) that an observation that is less than A or greater than B comes from the contaminating blunders. Since the distribution of the blunders (as defined by the Princeton study) was symmetric, you use those values (A and B) to define a trimmed average, which is an unbiased estimate of the mean you are trying to measure.

Charles Winsor (1895–1951), who had been on the Princeton faculty, had suggested a more sophisticated version of the trimmed average years before. Here is what Winsor suggested Instead of throwing out the values beyond the break-down point, you change all the values beyond the break-down point to the first observed value just inside the break-down point. This is called a **Winsorized average**, and the Princeton Robustness Study showed that it had a smaller variance than a trimmed average, so conclusions based on a Winsorized average are more precise.

Using mathematics far beyond the range of this book, the Princeton Robustness Study extended these findings into the estimation of multilinear regression. All of these findings are summed up in a book by Huber and Ronchetti [1]. Stephen Stigler of the University of Chicago [2,3] has explored some of the ramifications of robust statistical methods. In particular, in [3], Stigler re-examined the data used by the Cavendish committee (see Chapter 1) to estimate the distance from the Earth to the Sun. Applying these methods, he did no better than Cavendish with his careful selection of data.

10.1 SUMMARY

The Princeton Robustness Study was a large Monte Carlo study of the effects of contaminated distributions on the usual methods of statistical analysis used in science. The investigation was restricted to contamination (blunders) that had a symmetric distribution. This produced recommendations to use trimmed or Winsorized averages to estimate the center of a distribution. These recommendations were extended to more complicated models involving multilinear regression.

REFERENCES

1. Huber, P.J., and Ronchetti, E. (2009) *Robust Statistics*, 2nd Edition, New York, NY: John Wiley & Sons.
2. Stigler's description of modern robust procedures, http://home. uchicago.edu/~lhansen/changing_robustness_stigler.pdf (last modified June 2010).
3. Stigler, S. (1977) Do robust estimators work on real data? *Ann. Stat.*, 5, 1055–1098.
4. Kafadar, K. (2001) John Tukey and robustness, *Proceedings of the Annual Meeting of the American Statistical Association*, Aug 5–9, 2001, available at: http://www.amstat.org/sections/SRMS/ Proceedings/y2001/Proceed/00322.pdf
5. Alexander, M. (2009) Deconvolution problems in nonparametric statistics, Lecture Notes in Statistics (#193), Berlin: Springer-Verlag.
6. Joseph, H. (2016) Private communication.

CHAPTER 11

When the Blunder Is What You Want

IN CHAPTERS 9 AND 10, we considered the blunder as a nuisance, something that should not be in our data, and which has to be accounted for or eliminated from the calculations. But, there are times when it is the "blunder" that we want to examine, where we are interested in the contaminating distribution.

In the development of a new drug, the chemists or biologists will often produce a large number of candidates. These are organic chemicals or biological extracts that all should have a particular type of activity, like lowering blood pressure or interfering with a cytokine involved in the inflammatory process. Pharmaceutical companies have biologists and pharmacologists who can conduct tests of these compounds in vitro and in vivo. These tests can be used to eliminate many of the compounds that do not have adequate levels of the desired activity. But, there are usually several candidates that make it that far.

Then, the development begins to become more and more expensive as the new compounds are introduced into human clinical trials. The first phase (called Phase I) of studies (usually done in healthy volunteers) looks at each compound in terms of how the human body processes it, what doses are needed to achieve adequate blood levels, and whether there are any signs of trouble (like sudden rises in blood enzymes that suggest possible liver damage).

The second phase (Phase II) of human research has patients taking single doses or short sequences of doses to see if the compound seems to do some good and whether its metabolic fate is similar to that in healthy individuals. In the third phase (Phase III), large numbers of patients are treated in controlled randomized clinical trials.

At each step along the way, the cost of development gets higher and higher. The in vivo and in vitro studies are run within the facilities of the company or contracted out to research labs at costs calculated in the hundreds of thousands of dollars. The costs of the human studies start mounting into the millions. The management of the company has to decide whether to continue the development of a particular compound or to end its development and continue with another one in the series.

It has been my experience that the less knowledge we have about a compound, the better the compound looks. Once you go into the early human trials, problems arise. Some patients fail to respond. Was it because we used too low a dose, or is there a general lack of adequate efficacy? Would the next compound in the series (which has not yet entered human trials) be better?

Once a decision is made to engage in Phase III studies, the company is committing to costs of tens of millions of dollars. How can you look at the early trials and decide whether the current candidate is really having a therapeutic effect?

Seldom does a new drug have a beneficial effect on all patients. The test compound was constructed to have a specific pharmacological effect, blocking some natural process in the body. But, the human animal is the result of a long complicated evolutionary development, and there are usually multiple paths of enzymes or hormones that result in a given effect (such as swollen joints in rheumatoid arthritis). There will be some people, for whom the new compound "works" because, for these patients, their disease has resulted from problems with the biochemical pathway this compound affects. For others, alternative pathways take over, and the compound appears to have no effect on the disease.

When we put a group of patients into a clinical trial with the new drug, some of them will "respond," and some will not. We will have a contaminated distribution. The "blunders" are the patients who respond. If we have some measure of the disease (e.g., number of painful joints in rheumatoid arthritis), the distribution of the "blunders" is not symmetric as it was in the Princeton Robustness Study.

The questions faced by management of the drug company are, Have any patients responded? Is the measurement of efficacy from a contaminated distribution?

A key element in any statistical analysis is how well we can detect an effect. If we want to determine whether a new drug relieves pain, we don't try it out in only two or three patients. This is because any signal of efficacy will be lost in the random cloud of error uncertainty. How many patients should we use? There is a statistical calculation that answers that question. Using something called a power analysis, we can calculate how many patients we need to be reasonably sure of seeing an effect—if it is there.

It stands to reason that, if you are looking for something, it pays to have a good idea of where that might be. If you are looking for the car keys, you don't rummage in the garbage can or the linen closet. You have a better chance of finding them if you look, first, in the places you have put them before.

A similar idea occurs in statistics. You are using the data from a study to determine whether the true distribution of the measurement shows an effect. The more you can assume about the true distribution, the fewer patients you will need in order to have a good chance of seeing that effect. When the analysis is tailored to detect a well-defined probability distribution, it is called a **restricted test**.

It turns out that there is a restricted test [1] that can detect the presence of a small number of patients who come from another distribution, using fewer patients than would be needed if you did not use a restricted test.

With this restricted test, the drug company can determine whether the new compound has a useful effect on a subset of potential patients after a relatively small clinical trial.

11.1 SUMMARY

There are times when the "blunders," the contamination of a distribution, are what we are looking for. An example is shown of a problem in drug research, where using a model which includes a contaminating distribution of "responders" reduces the number of patients needed in a clinical trial to determine whether a new drug "works."

REFERENCE

1. Conover, W.J., and Salsburg, D.S. (1988) Locally most powerful tests for detecting treatment effects when only a subset of patients can be expected to "respond" to treatment, *Biometrics*, 44, 189–196.

CHAPTER 12

Parsing "Blunders"

IN THE CHAPTERS ON "Errors" (Chapters 2 through 10), we looked at the basic formula:

$$\text{Observation} = \text{model} + \text{error} \qquad (12.1)$$

where the model was a specific mathematical function involving variables that might influence the observation, and where the error term was a cloud of uncertainty (or Gamow's uncertain billiard ball). There are times, however, when we need to model the problem in terms of two or more different clouds of uncertainty. We saw in Chapter 11, how it is possible to test for the existence of the second source of uncertainty. Let's now look into the problem of dividing the uncertainty between its two clouds.

Suppose, we wish to survey the students in a high school to determine what percentage of them are engaging in some illegal or socially unacceptable behavior (like getting drunk or shoplifting). It would be difficult to get an honest count by just asking them. But, there is a technique known as the **randomized response survey** [1].

You choose a sample of students, as you would in any proper survey, and you approach each one with the following proposal:

> I am going to ask you two questions. One of them you may not want to answer. The other is innocuous. These two questions are (1) Did you engage in such-and-such an activity in the past month? and (2) were you born in this town? I want you to answer one of these questions with just a "yes" or a "no." I have a deck of cards here. You may examine them and shuffle them any way you want. Then, pull one out at random. Don't show it to me. If you pull out a queen, answer the second question. If you pull out any other card, answer the first question. Then, put the card back and reshuffle the deck.

All the surveyor learns from this student is "yes" or "no," because, at the time of the survey, there is no way of knowing which of the two questions has been addressed. How can we use this to find out how many of the students engaged in this forbidden activity? Let's set up a simple mathematical model:

$$\text{Prob}\left\{\text{answer is "yes"}\right\} = \pi p_1 + (1-\pi)p_2 \qquad (12.2)$$

where p_1 is the probability of being born in this town, p_2 is the probability of engaging in the forbidden activity, and π is the probability of drawing a queen from a deck of cards.

We know

$$\pi = 4/52 = 1/13, \text{ since there are four queens in the deck}$$

p_1 can be found by examining the vital records of the students in this school (how many were born in that town).

We can estimate the overall probability of getting a "yes" by counting the percentage of "yes" statements in the survey.

Suppose, as an example 43% of the students responded "yes" and 70% of the students in the school were born in that town.

Then Equation (12.2) becomes

$$0.43 = (1/13)(0.70) + (12/13)\, p_2$$

$$\text{or } p_2 = 0.408$$

Without our knowing the true behavior of any of the students surveyed, we can conclude that almost 41% of them are engaging in the illegal or socially unacceptable activity.

Randomized response surveys are an important part of the techniques used by survey statisticians. Survey methods form a major part of statistical activities, and the validity of the findings of a survey often depend on relatively simple equations like Equation (12.2).

Here is another example of a model where we seek to divide the data into those affected by error and the "blunders" from a contaminating distribution.

Methods have been developed for mapping the human genome and identifying the genes in a person's deoxyribonucleic acid (DNA). This opened the door to a new type of medical investigation. We know that many diseases have a genetic component. Some, like hemophilia, are clearly the result of inherited faulty genes. Others, like Alzheimer's disease, or cardiovascular disease, or breast cancer, are believed to be heavily influenced by faults in the DNA the patient inherited from his or her parents. Now, with the ability to map the genes and their possible mutations across the entire genome, it was felt that modern medicine could find the faulty genes, determine what protein those genes coded for, and devise a treatment that counteracted this genetic fault.

And so, medical scientists began to assemble groups of patients with a given disease and compare the genes in their DNA to a similar group of healthy subjects. Suppose we are looking for the genetic basis of cystic fibrosis. We get DNA samples from 100 cystic fibrosis patients and 100 age-matched healthy children and try to find which variant genes are correlated with having cystic fibrosis.

But, there is a problem. We have $N = 200$, the number of patients and subjects, and p is close to 40,000, the number of genes in human DNA. And, there may be more than 40,000 genes, since it is becoming clear that the expression of genes is governed by other parts of the DNA, the so-called epigenetic portion of the DNA, so p is much larger than even 40,000.

This is the "little N, big p" problem on steroids! And, the ghost of Carlo Bonferroni is sitting back and laughing at us. But, Bonferroni did not anticipate the cleverness of Yoav Benjamini and Yosef Hochberg of Tel Aviv University [2]. Here is their solution:

For each gene, we can measure some characteristic of its pattern of amino acids. These are usually in the form of single-nucleotide polymorphisms (SNPs) that describe the slight differences in DNA that occur when the nucleus of a cell is struck by cosmic rays. From this we can determine what percent of the sick children have a particular SNP and what percent of the healthy subjects have it.

We want to know how great that difference must be for us to consider that gene as possibly contributing to the disease. Let's call that measurement "t." The possible values of "t" come from one of two clouds of uncertainty. One cloud covers the situation where that particular gene has no role to play in the disease, so the average difference between the two groups of subjects has a mean of zero and is symmetrically distributed (just as we describe error in a regression problem). The other hypothesis is that this gene plays a role in the disease being investigated.

All but a very small portion of the genes play no role in the disease. Let's call that proportion θ. To put it another way, this is the probability that a gene selected at random is not involved in the disease. In keeping with convention, we write this probability as the Greek letter theta, because it is a parameter of the model and cannot be observed directly. The measurement we take, t, belongs to either the cloud of uncertainty associated with the hypothesis that it has no effect or to the cloud of uncertainty associated with it having an effect. As before, we are describing these clouds of uncertainty as probability distributions.

Consider t, the measure of the difference in this gene's characteristic between patients with the hereditary disease and the healthy controls. For any possible value of t, we can write the previous paragraph as an equation:

$$\text{Prob}\{t\} = \theta\,\text{Prob}\{t,\text{ when no effect}\}$$
$$+(1-\theta)\,\text{Prob}\{t,\text{ when there is an effect}\} \quad (12.3)$$

Let's select a value of t (say $t = 2.0$) that is relatively improbable, if there is no effect and more likely to result when this gene has an effect. If we use this value of t to select the candidate genes from the 40,000 in the genome, we will be mistaken some of the time, and a certain percentage of our "hits" will be false "discoveries." As an example, if we use $t = 2.0$, what is the percentage of "hits" that will be false? Benjamini and Hochberg called this percentage the **false discovery rate** (FDR), and here is how they estimated the FDR:

$$\text{FDR} = \text{Prob [“hit” given no effect]}/\text{Prob}\big[\text{“hit” in general}\big] \quad (12.4)$$

The numerator in Equation (12.4) is

$$\theta\,\text{Prob}\{t,\text{ when no effect}\}$$

But, θ, the proportion of genes that have no effect on the disease, is very large, so Benjamini and Hochberg let $\theta = 1$. Of course, θ is not equal to 1. If it were, then none of the genes are involved in the disease, and the whole exercise is futile. But, suppose there are only three genes involved, then θ equals 39,997/40,000, which is very close to 1.

I realize that mathematics is supposed to be precise, but this is a trick you often find in the mathematics of science. You replace a complicated expression by something that is reasonably close in value and much less difficult to deal with.

Now, Benjamini and Hochberg had to deal with the denominator of Equation (12.4). They could determine Prob{hit, given no effect}, but they still had to describe Prob{hit, given an effect}. Their solution was to estimate the denominator of (12.4) with

Total number of "hits"/total number of genes examined

With this neat combination of observation and theory, we can now choose a value of t that sets the FDR at whatever we want, say 20% or 10%. A complete discussion of the FDR approach to problems of small N and large p can be found in Reference [2].

12.1 SUMMARY

Two examples are shown of situations where we are dealing with a mixture of two distributions and wish to estimate aspects of the contaminating distribution. One of these is the randomized response survey, where the survey respondent is asked two questions with "yes" or "no" as the only possible answers. One question is innocuous; the other involves socially unacceptable behavior. The respondent is given a randomizing device, which is used to decide what question to answer. The theoretical probability of an "yes" answer to the innocuous question is known, as is the probability of choosing it. From this, we can deduce the incidence of the socially unacceptable behavior. The other problem deals with choosing from a very large number of variables using a relatively small number of observations. The FDR, proposed by Benjamini and Hochberg [2], provides a way of identifying the important variables with an estimable probability of error.

REFERENCES

1. Warner, S.L. (1965) Randomized response, a survey technique for eliminating evasive answer bias, *J. Amer. Stat. Assn.*, 60, 67–69.
2. Benjamini, Y., and Hochberg, Y. (1995) Controlling the false discovery rate: A practical and powerful approach to multiple testing, *J. Roy. Stat. Soc.*, Ser. B, 57, 289–300.

IV

Lies

CHAPTER 13

The Reigns of Kings

For God's sake, let us sit upon the ground,
And tell sad stories of the death of kings;
How some have been depos'd; some slain in war;
Some haunted by the ghosts they have depos'd;
Some poison'd by their wives; some in sleep kill'd;
All murder'd.

IN SHAKESPEARE'S PLAY, *Richard II*, this is King Richard's mournful take on the lives of kings as he is being deposed and about to die. This raises the question of how long do kings live? According to Roman legend, Romulus, the first king of Rome reigned for 38 years, and his son reigned for 42 years. Can this be true?

Table 13.1 displays the lengths of the legendary reigns of the kings of Rome. Table 13.2 displays the lengths of reigns of the kings of England from William I to Richard II.

I have chosen to compare these two sequences because we know that the reigns of early English kings are accurate and the conditions of life and medicine were very similar during the two periods of time, so the ordinary force of mortality was the same.

The first thing that is obvious is that the reigns of English kings show much more variability than the reigns of the Roman kings.

TABLE 13.1   Legendary Reigns of the Early Kings of
Rome, an Example of Falsified Data and Its Frequent
Lack of Adequate Variability

Early Kings of Rome	Length of Reign (years)
Romulus	38
Nuna Pompilius	42
Tulus Hostilius	32
Cinius Marcius	26
Lucus Tarquinius Priscus	38
Servius Tulius	34
Lucus Tarquinius Superbus	25

TABLE 13.2   Reigns of the First 13 Rulers of England,
with Much Greater Variability than the Legendary
Reigns of Roman Kings

Kings of England	Length of Reign (years)
William I	12
William II	14
Henry I	36
Stephen	20
Queen Maud	1
Henry II	35
Richard I	11
John	17
Henry III	57
Edward I	37
Edward II	21
Edward III	50
Richard II	23

The lack of variability is often a hallmark of faked data. In real
life, every king does not rule for a long and prosperous reign. As
Shakespeare knew, some kings are killed by enemies or in battle
or are overthrown. It is not uncommon in real life to find kings
who ruled for a year or less. Considering the lifespans of men
before the rise of modern medicine, it is highly improbable that
a king who starts his reign in his 40s would live for more than

10 or 15 years. So, the reigns of the Roman kings are of questionable validity both in terms of length and variability.

There is a statistical test for comparative variability. Variability is measured by the variance of a set of data. An estimate of the variance of the list in Table 13.1, Roman kings, is 40.6. An estimate of the variance of the list in Table 13.2, English kings, is 223.8. If the two sets of data were drawn at random from sets of data with the same underlying variance, then the ratio of those variance estimates should range between 0.25 and 2.5. The ratio for the variance in Table 13.2 to that of Table 13.1 is 5.10. If they were representative of the same process, then we have observed something that is highly improbable (probability less than 6/1000). It is reasonable to reject these legendary reigns as faked data.

If the Romans were telling lies, how about the ancient Hebrews? Table 13.3 displays the length of reigns of the kings of Judah, taken from the Biblical *Book of Kings II*. These numbers look more real than the Roman kings. There is a great deal of variability. Two kings reigned for more than 50 years, three for less than 10. The estimated variance for these reigns is 313.9, larger than for the English kings, but the ratio of the two variance estimates equals 1.402. The probability of getting that ratio when both sets of data come from the same distribution is more than 25%.

Thus, the statistical examination shows that the Romans were liars but the Hebrews were not.

The failure of faked data to have sufficient variability holds as long as the liar does not know this. If the liar knows this, his best approach is to start with real data and use it cleverly to adapt it to his needs.

An example of such sophisticated lying occurred in the work of Sir Cyril Burt (1883–1971) a British psychologist from the first half of the twentieth century (see Reference [1]). Burt made his career from following a group of identical twins, who had been separated at birth. His studies of their achievements on standardized tests as they grew up showed, conclusively, that nature was more important than nurture. In spite of being raised in different

TABLE 13.3   Reigns of the Kings of Judah from
the Bible—Does This Show Adequate Variability?

Kings of Judah	Length of Reign (years)
Rehoboam	17
Abijah	3
Asa	41
Jehoshaphat	25
Jehoram	8
Ahaziah	1
Athaliah	6
Joash	40
Amaziah	29
Uzziah	52
Jotham	16
Ahaz	17
Hezekiah	29
Manasseh	55
Amon	2
Josiah	31
Jehoahaz	3 months
Johiakim	11
Johiachin	3 months
Zedekiah	11

homes (sometimes in different countries), the grades on standardized tests of any two twins were consistently closer than might
be expected from children raised in different environments. As a
result of Burt's studies, the British government established standardized tests to be given to children at age 11, which were used
to designate which of the children went on to further academic
training and which of the children were given manual training.

After Burt's death in 1971, an attempt was made to contact some
of his coauthors to verify some questions. His papers always had
female coauthors, and everyone assumed that these were women
who helped in the testing and tabulation, but none of them could
be found.

Although there is still some dispute over how much, if any, of Burt's published work was faked, the general conclusion of his critics is that none of these coauthors existed. His critics claim that Burt had been using faked names for coauthors throughout his career. His critics insist that he had not followed any identical twins separated at birth and that all of his data were faked.

But, the data he reported in his papers looked authentic. It had adequate variability. It met various tests for randomness, although the same sets of numbers sometimes repeated themselves in different papers describing different situations. Then, his investigators found an almanac in his library, where many pages had been marked. Burt had copied numbers out of that almanac, moving decimal places to keep his numbers in the right range.

It has been my experience that most data fakers (whom we have identified) are not as sophisticated as was Cecil Burt, and the lack of adequate variability remains the major mark of lying data.

13.1 SUMMARY

Data that have been faked usually lack the variability of real data. When the reputed lengths of reigns of the ancient kings of Rome are compared to the lengths of reigns of kings of England, the reigns of English kings are much more variable. The reigns of the kings of ancient Judah, as described in the Book of Kings II, meet this test of variability. However, faked data may have adequate variability if the faker is sophisticated and starts with real data, as did Cecil Burt.

REFERENCE

1. Samelson, F. (1997) What to do about fraud charges in science; or, will the Burt affair ever end? *Genetica*, 99(2–3), 145–151.

CHAPTER 14

Searching for the "Real" Davy Crockett

D AVY CROCKETT (FIGURE 14.1), the hero of a television series
and a movie in the 1950s, was a real person. He was a member
of the United States Congress in the 1830s. A faction of the Whig
Party had tried to nominate him as a candidate for President. He
also wrote two or three books. Two or three? That is a question
that this chapter will explore.

Crockett was elected to Congress from Tennessee in 1832
as a supporter of President Andrew Jackson, who had been his
commander when he fought in the Indian Wars. When he ran
for re-election in 1834, he published a campaign autobiography
entitled, *A Narrative of the Life of David Crockett*. In the autobiography, he wrote that he had attended school for only 3 days in
his life and that he was illiterate until he was in his 30s, when his
wife taught him to read and write. The style of his writing can be
seen in his preface:

> I don't know of anything in my book to be criticized on
> by honorable men. Is it on my spelling ?—that's not my
> trade. Is it on my grammar?—I hadn't time to learn it, and

FIGURE 14.1 David Crockett (1786–1836) woodsman, member of Congress, Presidential candidate, fighter for Texas, author—how many books did he really write? (Courtesy of Shutterstock.com.)

make no pretensions to it. Is it on the order and arrangement of my book?—I never wrote one before, and never read very many, and, of course, know mightly little about that. Will it be on the authorship of the book—this I claim, and I'll hang on to it like a wax plaster. The whole book is my own, and every sentiment and sentence in it. I would not be such a fool, or knave either, as to deny that I had it run hastily over by a friend or so, and that some little alterations have been made in the spelling and grammar.

During his second term in Congress, Crockett broke with his hero, Andrew Jackson. He opposed Jackson's plans to forcibly remove the five "civilized" Indian tribes from their well-watered homelands in North Carolina, Tennessee, Georgia, Alabama, and Florida and transport them—men, women, and children—to the arid lands of what was later to become the state of Oklahoma. This is the infamous "Trail of Tears" that has been considered one of the great injustices in the relation between the U.S. government and the Native Americans.

Not only did he break with Jackson on the Trail of Tears, but he also opposed the president when Jackson destroyed the national bank. His sympathies and his politics were sliding into that of the Northeastern Whigs, who supported a strong currency and a tariff to protect home-grown industries. After his defeat for a third term in Congress, he was invited by a group of New England manufacturers to go on an extended tour of the Northeast.

The result of that trip was another book, supposedly written by Davy Crockett, *An Account of Col. Crockett's Tour to the North and Down East*. Was this book really written by Davy Crockett? Or was it the product of a publicity agent for the manufacturers? In his book, Crockett described the idyllic conditions of the young women working in the New England factories, where they were carefully guarded and locked up at night in their dormitories to protect their virtue. He does not mention the terrible working conditions but praises how the girls were able to send money home to their families (most of them farmers).

Crockett was elected to one more term in Congress. At the end of this service, he was 50 years old, and he set off to aid the Texans in their revolt against Mexico. He arrived in San Antonio and was among the defenders of the Alamo when it fell to the Mexican dictator, Santa Anna. He either died in the fight or was executed by Santa Anna's forces after the surrender of the Alamo.

Later that year, a first person narrative of his journey to Texas and the battle for the Alamo was published: *Col. Crockett's Exploits and Adventures in Texas*. When I visited the Alamo, I found that pages from Crockett's book (in the form of a diary) had been framed on the walls of the shrine, depicting the final hours of the battle. Did Crockett, in fact, write this final book, or was it the product of some unknown literary hack, who was trying to make some money on Crockett's reputation?

To let the reader judge whether these were books written by the same man, here are some samples. From *The Narrative of the Life of Davy Crockett*, here is the famous scene of his fight with a bear:

> When I got there, they had treed the bear in a large forked poplar, and it was setting in the fork. I could see the lump, but not plain enough to shoot with any certainty, as there was no moonlight … At last I thought I could shoot by guess and kill him; so I pointed as near the lump as I could and fired away. But the bear didn't come, he only clumb up higher, and got out on a limb, which helped me to see him better. I now loaded up again and fired, but this time he didn't move at all. I commenced loading for a third fire, but the first thing I knowed, the bear was down among my dogs, and they were fighting all around me. I had my big butcher in my belt, and I had a pair of dressed buckskin breeches on. So I took out my knife, and stood, determined, if he should get hold of me, to defend myself in the best way I could…

Here is a sample from the *Col. Crockett's Tour of the North and Down East*:

> We often wonder how things are made so cheap among the Yankees. Come here and you will see women doing men's work, and happy and cheerful as the day is long, and why not? Is it not better for themselves and families, instead of sitting up all day busy about nothing? It ain't hard work, neither, and looked as queer to me as it would to one of my countrywomen to see a man milking the cow, as they do here.

Here is a sample from the *Texas* book describing his fight with a cougar, which is reminiscent of the fight with a bear in his first book:

> …there was no retreat either for me or the cougar, so I leveled my Betsey and blazed away. The report was followed by a furious growl, (which is sometimes the case in Congress,) and the next moment, when I expected to find the tarnal critter struggling with death, I beheld him shaking his head as if nothing more than a bee had stung him. The ball had struck him on the forehead and glanced off, doing no other injury than stunning him for an instant, and tearing off the skin, which tended to infuriate him the more. The cougar wasn't long in making up his mind what to do, nor was I neither, but he would have it all his own way, and vetoed my motion to back up. I had not retreated three steps when he sprang at me like a steamboat: I stepped aside, and as he lit upon the ground, I struck him violently with the barrel of my rifle, but he didn't mind that, but wheeled around and made at me again…I drew my hunting knife, for I knew we would come to close quarters before the fight would be over.

What else do we have of Crockett's purported writings? There were his speeches in Congress. At that time, there was no *Congressional Record* to archive the speeches of Congressmen verbatim. Most of the information we have about Crockett's speeches are in the form of reporter's notes or paraphrases. However, it is possible to find small parts of the speeches that are purported to be verbatim.

So, how can we tell which of these books (and speeches) are, in fact, written by the legendary Davy Crockett? Frederick Mosteller (1916–2006) and David Wallace (currently professor emeritus, University of Chicago) were the first to look at word counts to distinguish between different authors [1]. The basic idea is that, when we use language to convey something, it is necessary to link the important words with prepositions, adverbs, and conjunctions in order for the sentence to make sense. Mosteller and Wallace called these "noncontextual" words.

Table 14.1 shows the list that they compiled of noncontextual words. The use of these noncontextual words is an essential part of a person's speaking or writing pattern. They are inserted without the user thinking about them, since they are the glue that puts ideas together in a way that the speaker or writer finds best.

Mosteller and Wallace started with a compilation of the frequency of common words in the English Language that had been published in 1958. They examined the frequencies with which those words occurred in known writings of Madison and Hamilton. They also counted frequencies of usage in the novel,

TABLE 14.1   Noncontextual Words, Whose Frequencies Can Be Used to Identify the Author of a Given Passage, Since the Use of These Connecting Words Reflect the Author's Unconscious Use of Grammatical Structures

Mosteller and Wallace's Noncontextual Words	
upon	commonly
also	consequently
an	considerable
by	according
of	apt
on	direction
there	innovation
this	language
to	vigor
although	kind
both	matter
enough	particularly
while	probability
whilst	works
always	though

Ulysses, by James Joyce. The list they put together consisted of words that met two criteria:

1. The frequency with which a single author used the word did not change from subject to subject or over the course of time.

2. The frequencies of use of each of these words differed from author to author.

The frequency of usage of any one noncontextual word was not sufficient to distinguish between two authors. For instance, Mosteller and Wallace used this tool to distinguish between the Federalist papers that were written by Madison and Hamilton. In their initial examination of the papers, they found that the word "whilst" was used by Madison and "while" by Hamilton in their identified writings. However, this was not sufficient to distinguish which man wrote which paper since the word appeared only seldom, and, in many of the disputed papers, it did not appear at all. However, it is possible to combine the frequencies of many such words into a single statistical model and use that model to distinguish between authors.

My daughter, Dena Salsburg Vogel, and I went through the books and speeches attributed to Davy Crockett and identified nine noncontextual words which appeared often enough to be used in a statistical analysis of the purported works of Davy Crockett. A full report of our project can be found in Reference [2]. Table 14.2 describes the frequencies of the nine words as they appear in the *Narrative*, the *Tour*, the *Texas* book, and his purported speeches. It can be seen that the *Texas* book differs from the other three in its usage of the words "an," "this," "to," and "though."

There is a test to determine if two counts of a given word have the same probability of occurring. When it is applied to the data in Table 14.2, the usage of these four words marks the *Texas* book as different from the others, but there are no significant differences in frequencies of any of these words among the other three sources.

TABLE 14.2   Frequencies of Most Frequent Noncontextual Words Used in the
Three Books Attributed to Davy Crockett

	Frequency/1000 Words			
Noncontextual Word	The *Narrative*	The *Tour*	The *Texas*	Speeches
also	0.75	0.32	0.41	0.51
an	2.63	2.21	**4.93**	2.54
there	3.75	4.90	3.43	1.52
this	4.75	5.22	**2.60**	7.10
to	36.79	33.21	**28.78**	30.93
both	1.00	0.47	0.14	0.00
while	1.13	0.63	0.68	2.54
always	0.13	0.79	0.55	0.00
though	0.25	0.00	**1.51**	0.00

Boldface numbers indicate where the *Texas* differs significantly from the others.

Thus, Davy Crockett died or was executed at the Alamo, but, in spite of the framed diary pages on the walls of the shrine, he left no record of his final journey.

When we submitted a paper describing our findings to the journal, *Chance*, the editor wanted to know if we could find out who did write the *Texas* book. There was a lot of book writing (some of it good and some awful) during the 1840s. Most of these have been destroyed or are hiding in the back shelves of obscure libraries. So, we could not expect to answer that question if the author was a scrivener of popular potboilers. However, we tried. We compared the *Texas* book to works by the two leading contemporary authors, James Fenimore Cooper and Nathaniel Hawthorne. We could show conclusively that Cooper and Hawthorne were not the same people. We also had to reject the hypothesis that either of them wrote Crockett's *Texas* book (see Table 14.3).

14.1 SUMMARY

Davy Crockett is purported to have written three books. One of them was an autobiography written to support his election to Congress in 1834. The second was a description of his tour of manufacturing plants in New England, and the third was a purported

TABLE 14.3   Frequencies of Noncontextual Words in the Book *Texas* Attributed to Crockett and Works by Cooper and Hawthorne

Noncontextual Word	Frequencies/1000 words		
	The *Texas*	Work by Cooper	Work by Hawthorne
upon	2.47	**0.22**	1.95
also	0.41	0.22	0.98
an	4.93	2.69	3.90
by	3.70	**7.17**	5.85
of	30.29	**44.11**	**62.90**
on	5.07	6.94	4.88
there	3.43	**1.34**	3.90
this	2.60	**3.05**	**7.31**
to	27.78	27.54	29.74
both	0.14	**2.24**	0.49
while	0.69	0.45	0.49
always	0.55	0.22	1.46
though	1.51	1.34	**4.88**

Bold face numbers indicate where the Texas differs significantly from Cooper or Hawthorne.

I have applied this method of analysis to the books of the *Hebrew Bible*, and the results can be found in Reference [3].

diary written on his way to Texas and death in the battle for the Alamo. The last book, the one describing his final trip, has long been suspect among historians. A statistical method based on counts of noncontextual words (which was pioneered by Mosteller and Wallace) was used to compare the three books and Crockett's speeches in Congress. The *Texas* book is shown to be a forgery.

REFERENCES

1. Mosteller, F., and Wallace, D.L. (1984) *Applied Bayesian and Classical Inference: The Case of the Federalist Papers*, 2nd Edition, New York, NY: Springer-Verlag.
2. Salsburg, D., and Salsburg, D. (1999) Searching for the real Davy Crockett, *Chance*, 12(2), 29–34.
3. Salsburg, D. (2013) *Jonah in the Garden of Eden: A Statistical Analysis of the Authorship of the Books of the Hebrew Bible*, available at: amazon.com (as an e-book).

CHAPTER 15

Detecting Falsified Counts

THE BOOKS OF THE Hebrew Bible (the "Old Testament") were written and assembled over a period of about 2200 years. Starting before the destruction of Herod's temple by the Romans in 72 of the Common Era and extending for about 200 years afterwards, the rabbis of the Talmudic period of Judaism organized and canonized the books that eventually became our current version of the Hebrew Bible.

The rabbis were confronted with a large number of books and several different versions of some books. They left no clear description of the process of canonization, but there are statements in the Talmud that indicate why some books were left out of the canon, usually because they did not trust that the version they had on hand was an accurate version of the original or because the only versions they had on hand were in Aramaic rather than Hebrew. There are suggestions that some books were left out because the rabbis were suspicious of their provenance.

An early version of the Hebrew Bible was translated into Greek around 250 years before the Common Era. This version is called the Septuagint because it was supposedly translated by a group of 70 scholars and was written for the Jewish community in Egypt, many of whom could not speak or read Hebrew.

There are 13 books in the Septuagint that were left out of the Hebrew Bible in its final canonization. In addition to these 13 books, there is also an extensive addition to the Biblical Book of Esther that was in the Septuagint but not in the final canonized Bible. We can only surmise why these books were not included in the final version. My own reading of the few passages in the Talmud that refer to the canonization suggest to me that the rabbis of the Talmud were aware that the accumulated literature referring to earlier days contained some forgeries. To determine the authenticity of a given book, they appear to have relied on earlier references to that book and inconsistencies between the questionable book and others. For instance, the Book of the prophet Ezekiel was almost left out because the prophet's description of ritual sacrifices in the Temple, when it would be restored, disagreed with the details of sacrifice described in the Book of Leviticus. The reference is to a single rabbi who "sat up all night" and used several pots of ink finding slight twists of language that enabled him to reconcile the two descriptions.

The books that appear in the Septuagint but not in the final version of the Hebrew Bible have been collected as the Apocrypha, which is considered part of the Holy Bible by some branches of the Orthodox Christian Church.

Two of the books in the Apocrypha are attributed to the prophet Esdras, whom most Biblical scholars identify as being the same person as the prophet Ezra, whose book remains in the canon. The two books of Esdras contain material found in the canonized books of Ezra, Nehemiah, and Chronicles, but there were apparently other reasons why the Talmudic rabbis rejected them.

In the Book of Esdras I, Chapter 5, Verses 1–43, there is a detailed description of the Jews who returned from the Babylonian exile to rebuild the Temple in Jerusalem. It describes the return of 43 groups or families with exact counts of how many people were in each group. It also lists the number of horses, camels, mules, and asses. In all, a collection of 47 numbers. These numbers are all exact. With phrases like

...The descendants of Binnui 648; The descendants of Bebai 623; the descendants of Agad 1322...

Table 15.1 displays the numbers of returnees attributed to each family from the Book of Esdras I. Table 15.2 displays the counts of unit digits. Notice the large number of times a number ends in "2" and that an ending of "0" occurs only once.

TABLE 15.1 Counts of Groups of People Returning to Build the Temple in Jerusalem According to Esdras I

Counts of Returnees According to Esdras I			
Parosh	2172	Bethasmoth	42
Shephaitah	472	Kirjath-jearim	25
Arah	756	Chephirah	743
Pahath-moab	2812	Chadiasans	422
Elam	1254	Ramah	622
Zattu	945	Michmach	122
Chotbe	705	Bethel-Ai	52
Bibbui	648	Magbish	156
Bebai	623	Elam	725
Azgad	1322	Jericho	345
Adonikam	667	Senaah	3330
Bigvai	2066	Jedaiah	972
Adin	454	Immer	2052
Ater	92	Pashur	1247
Kolan	77	Harim	1017
Azuru	432	Jeshua	74
Annias	101	sacred singers	128
Bezai	323	doorkeepers	139
Jarah	112	temple slaves	372
Baiterus	3005	Deliliah	652
Bethlehem	123	Camels	435
Netophah	55	Horses	7036
Anathoth	158	Mules	245
		Asses	5525

TABLE 15.2   Counts of Units Digits from Table 15.1

Counts of Units Digits According to Esdras I		
Unit Digit	Expected Count[a]	Observed
0	4.7	1
1	4.7	1
2	4.7	16
3	4.7	4
4	4.7	3
5	4.7	10
6	4.7	4
7	4.7	4
8	4.7	2
9	4.7	1
Probability		0.0003[*]

If the data were real, the unit digits should tend to be uniformly distributed among all digits from 0 to 9.

[a] Expected count if all digits were equally probable.

[*] Probability that this pattern would occur at random.

When someone sits down to write numbers out of his or her head, no matter how haphazard or "random" that person thinks he or she is doing it, there is going to be a psychological phenomenon known as "digit preference." Everybody has a subconscious tendency to favor some digits over others. If, in fact, the numbers quoted in the book of Esdras were taken from a real-life census, then the least significant figure, the units digit, should be completely random. Every possible digit from 0 to 9 should be equally probable. This leads to a technique used at the U.S. National Institutes of Health (NIH) to flag possible faked data.

There are 47 numbers in Esdras. In 16 of those, the final digit is "2." In 10 of them, the final digit is "5." There is only one "1" and only one "9." Had all possible digits been equally probable, we should have had between three and six occurrences of each digit. If the data came from true counts, then we would expect to see this great a difference from those expected frequencies less 0.04% of the time. If anyone had turned in a report using these numbers to the NIH, it would have been declared fraudulent.

Why would the author of the book of Esdras go to the trouble of presenting us with such exact numbers? The fraud would not have been detectable if he had written, "Of the descendants of Binnui approximately 650, of the descendants of Bebai, around 600, of the descendants of Agad over 1300…"

The British soldier of fortune, T.E. Lawrence, has a possible answer in his book, *The Seven Pillars of Wisdom*. During World War I, Lawrence was traveling through the Middle East, seeking to raise local troops to join the allies and fight the Ottoman Empire. He would often meet with a desert sheik, and the sheik would say he can produce 426 armed men, but when it came time to collect these soldiers, 6 or 7 men with old rifles would show up.

Was the sheik lying? At first Lawrence thought so and wondered why so many of his potential allies fell short. Then, he realized that the sheik was not purposely lying. In that culture, numbers did not have the exact meaning that is needed when we use them in the modern scientific world. In that culture, numbers were a form of flowery speech—the more exact the number seemed, the more flowery. Perhaps, the culture that T.E. Lawrence had to deal with was the same culture that had the author of the book of Esdras describe the families that returned to Jerusalem with exact, although purely fictitious, numbers.

Let's examine some numbers that appear in the canonized books of the Hebrew Bible. In the *Book of Numbers* (verse 1: 1–47 and verse 26: 1–63), Moses and Aaron take a census of fighting men among the Israelites. They take this census twice, once upon leaving Egypt and once just before entering the Holy Land. They report the number of fighting men "from 20 years old and upward." In both censuses, they count around 600,000 fighting men. Since there are usually at least as many women as men in any population, and this figure does not include men and boys under age 20, these 600,000 fighting men translates to more than 1.5 million Israelites who were liberated from Egypt.

The increase in living populations, whether they be bacteria, foxes, or humans, follows a mathematical model known as exponential growth. If we look at those times when the world human population could be effectively estimated, we can plot that growth back to the time of Moses and the liberation from Egypt. The best estimate for the entire worldwide human population at the time of the Exodus is about 50 million.

One and a half million Israelites enslaved in Egypt at a time when the best estimates of the total number of humans on Earth were about 50 million seems incredible. These 50 million were scattered across the Earth, from Asia to Europe, from the Pacific to the Atlantic coast of Africa. It would be an overestimate to think that 1 in 50 lived in Egypt. But, even with that exaggerated amount, it means that there were 1.5 million Israelites enslaved by one million Egyptians, a highly improbable situation. How valid are these census counts from the *Book of Numbers*? [1]. Table 15.3 displays the counts of thousands and hundreds digits from these censuses.

TABLE 15.3   Counts of Digits from the 100s and 1000s Places in the Two Censuses of Fighting Men among the Israelites in the *Book of Numbers*.

Digit Counts from Censuses of Soldiers—*Book of Numbers*			
Digit	Expected Count[a]	Hundreds	Thousands
0	2.6	1	1
1	2.6	1	2
2	2.6	4	5
3	2.6	1	4
4	2.6	6	2
5	2.6	6	2
6	2.6	5	3
7	2.6	2	3
8	2.6	0	3
9	2.6	0	1
Probability		0.028[*]	0.852[*]

The 100s digits appear to be faked, but the 1000s digits are not.
[a] Expected count if all digits were equally probable.
[*] Probability this pattern would occur at random.

The number of fighting men in each tribe is reported to the nearest hundred (with a few 50s thrown in here and there). If we concentrate on the hundreds digit, we see that, of 26 numbers in the two censuses combined, none of the hundreds digit is an "8" or a "9," and there are six counts of both "4" and "5." These discrepancies from the equal numbers that would have been expected have about a 3% chance of occurring at random.

However, the thousands digits tell a different story. Every digit is represented and none more often than five times. This kind of pattern is well within what one might expect normally if each of the digits was equally probable.

So, what does this mean? The Hebrew word, "elef," which we translate as "thousand," did not always mean the exact number 1000. The prophet known as the Second Isaiah, who lived during the sixth century b.c.e., around 1000 years after the Exodus from Egypt, used the word "elef" to refer to a family grouping of indefinite size (see Isaiah 60:22). There is good reason to believe that it also once referred to a fighting unit. If, in these ancient armies, a fighting unit was like a squad, the minimal size fighting unit in modern armies, then the census did not count 600,000 soldiers but 600 fighting units, possibly as low as 5–10 men each. Thus, the 600 "elifim" might easily have been 3–6,000 fighting men—still a very formidable force. But, this brings us down to 12,000–13,000 Israelites liberated from Egypt.

Where did the hundreds digits, the faked ones, come from? The initial *Book of Numbers* was probably composed before the time of writing and was passed on as an oral tradition. After no more than 100 years of oral transmission, it began to be written down. But, parchment and papyrus don't last forever, and after 20 years or so of constant use, the book would have to be re-copied, and re-copied, again and again. In this copying, the letters were, first, a version of Phoenician and later were taken from the Aramaic alphabet. Imagine a scribe trying to copy these thousands of words from a torn old manuscript, dipping his pen into ink for each stroke. Errors creep in. Letters or even whole words get inserted in nonsensical places or get left out. If, in their earlier versions, these were not considered the immutable Word of the Lord, there might have been a temptation for one of the scribes to "improve" upon the manuscript, perhaps engage in some flowery language by adding in the hundreds digits, because, by that time, "elef" had lost its original meaning as a fighting unit.

Digit preference as a hallmark of faked data is not restricted to traditional ancient works. As shown in Reference [2], census reports from third-world countries often show strong digit preferences that suggest that many of the numbers have been drawn up at a bureaucrat's desk rather than actually counted in the field. The question of the validity of some African censuses is reexamined in Chapter 17.

15.1 SUMMARY

The Book of Esdras I in the Apocrypha provides an example of faked data in the lists of the numbers of each family unit who returned to Jerusalem to rebuild the Temple after the Babylonian Exile. Digit preference is an unconscious psychological characteristic. Asked to put down a sequence of numbers "at random," almost everyone will show a predilection to avoid some digits and repeat some other ones. The counts of returnees from Esdras I show a clear example of digit preference. Another set of numbers in the Bible occurs in the *Book of Numbers* in the censuses taken of the number of fighting men among the Israelites leaving Egypt. The numbers are reported to the nearest 100, and the hundreds digits display digit preference. However, the thousands digits do not. Since one of the original meanings of the Hebrew word, "elef," which we translate as 1000, also meant a fighting unit (which could have as few as 5 men), the finding that the thousands digit is properly random suggests that someone, in the copying and recopying of the text over 2000 years, decided to make the numbers more "exact" and added in the hundreds digits.

REFERENCES

1 Salsburg, D. (1997) Digit preference in the Bible. *Chance*, 10(4), 46–48.

2 Nagi, M.H., Stockwell, E.G., and Snavley, L.M. (1973) Digit preference and avoidance in the age statistics of some recent African censuses: Some patterns and correlates. *Int. Stat. Rev.*, 41, 165–174.

CHAPTER 16

Uncovering Secrets

O N FEBRUARY 12, 1941, a fleet of ships crossed the Mediterranean Sea from Italy to bring a German panzer division to the aid of the Italian army fighting against the British in Libya. Under the command of General Erwin Rommel, the 15th Panzer Division brought with them a newly developed light German tank specifically designed for desert warfare. The British were soon able to impose a blockade across the Mediterranean and block further reinforcement of the Axis army, but the new German tanks were already there (Figure 16.1).

In the initial skirmishes, the British soon learned that this light desert-perfect German tank was vastly superior to anything they had. They realized that they would probably lose any massive tank battle that might ensue. So, they retreated from Libya and into fortified positions on the Egyptian border. They began a war of attrition, using airpower and lightning attacks from hidden bases deep in the Libyan Desert, trying to knock out as many of the new German tanks as they could. If they could cut down enough of the German tanks, they could then attack in full and win the resulting tank battle.

But, how many of these superior desert tanks did Rommel have? At what point would the attrition be sufficient for the British to win an all-out battle?

FIGURE 16.1 German General Erwin Rommel (1891–1944) directing the battle in North Africa. How many tanks did Rommel have? (Courtesy of Shutterstock.)

The number of tanks the German army brought over was a military secret, but the analysts on the British side knew that the factories that produced those tanks had given them a consecutive sequence of serial numbers. Could they use the serial numbers of the tanks they managed to knock out and determine the total number of tanks produced (since almost the entire production of these tanks had been sent to Africa)? While British spies sought to find the number of tanks produced by probing their sources within Germany, the analysts lined up the serial numbers of tanks destroyed so far and used a statistical model.

If you have a collection of orderly numbers and you draw at random from that collection, there is a consistent estimator of the maximum and of the minimum value in that complete collection. Gottfried Noether (1915–1991) described this method (and the story of Rommel's tanks) in his 1976 textbook on nonparametric statistics [1].

In his example, Noether proposed that if someone is trying to flag a taxi, and every taxi that passes appears to be in use, from the taxi numbers, can you figure out how many taxis there are in the city? Suppose you observe the following numbers:

322, 115, 16, 98, 255, 38

Order the series:

$$16, 38, 98, 115, 255, 322$$

Examine the gaps between numbers:

$$38 - 16 = 22$$
$$98 - 38 = 60$$
$$115 - 98 = 17$$
$$255 - 115 = 140$$
$$322 - 255 = 67$$

Take the median of those gaps (60 = 98 − 38).

A consistent estimator of the maximum number of taxis is the largest number observed plus the median gap (382 = 322 + 60).

Using this method, the British army analysts estimated the maximum and minimum serial numbers of the German tanks, and got an upper bound on the total number of tanks Rommel had under his command. Thus did the structured careful ordering of the serial numbers of the German factory tanks reveal a closely held military secret.

Abraham Lincoln may have freed the slaves in the United States, but slavery continued and continues throughout the world, even into the twenty-first century. In the period from 2000 to 2012, there were approximately 2500 different reports or surveys of examples of forced labor in different countries. In 2012, the International Labor Organization (ILO) tried to use these surveys to determine the total number of slaves in the world.

There were two problems with these surveys. They did not necessarily include all the slaves, and there was some overlap with as many as 46 names appearing in four or more of the reports. The analysts at the ILO were able to use the statistical methods of ecology to solve this problem and estimate that 20.9 million people are held as slaves or under forced labor throughout the world.

Consider the problem of determining the number of a species of wild animals (say foxes) in a given area. Since it was first proposed in

1959, ecologists have been using capture–recapture techniques for those counts. In its simplest form, you capture a group of animals during a particular time of day and you mark and release them. You return again at the same time of day and capture another group of animals. Suppose, you capture and mark 50 animals on the first day and, on the second day, 10 of the 50 animals you trap are already marked. To get a rough estimate of the total number of foxes in the area, you note that in your second capture, one-fifth of the animals had been captured before, therefore you estimate that the probability of capture is 20%, that your initial capture contained 20% of the total number of animals, so you estimate that the total number is 250 (= 5 × 50).

Since the initial papers in the 1950s, capture–recapture techniques have become quite sophisticated. The statistical models include parameters that describe the birth rate and death rate among the animals being counted. Capture–recapture techniques include models for many capture sessions over a period of time, which allow ecologists to use indirect markers like scat or cameras triggered by animals taking bait, or collections of insects.

The ILO used one of the more sophisticated capture–recapture models where you do a number of captures and look at the frequencies at which the same names of slaves appear on multiple reports. You, then, have a list of the number of names that appear on only one list, on two lists, on three lists, and so on. You are missing the number of names that appear on no list, but that can be estimated from the numbers of multiple appearances just as the total number of foxes can be estimated from two capture–recapture sessions. See Reference [2] for a complete description of the methodology and Reference [3] for the ILO report.

16.1 SUMMARY

When there are regularities in data that can be detected by statistical methods, it is possible to uncover secrets. Statistical techniques were used to determine the number of tanks in the German Africa Corps on the basis of serial numbers of captured tanks. Capture–recapture techniques from ecology were used by the ILO to estimate the number of slaves being held in the world in the twenty-first century.

REFERENCES

1. Noether, G.E. (1990) *Introduction to Statistics: The Nonparametric Way.* New York, NY: Springer Tests in Statistics.
2. Bøhning, D. (2016) Ratio plot and ratio regression with applications to social and medical sciences. *Stat. Sci.*, 31(2), 205–218.
3. ILO (2012) *Global Estimate of Forced Labor: Results and Methodology*, Geneva: International Labor Organization.

CHAPTER 17

Errors, Blunders, or Curbstoning?

S UPPOSE YOU ARE A young recent high school graduate in a third-world country where the chronic unemployment rate for high school graduates is 60%. Your uncle works at the census bureau for the government, and he gets you a temporary job as a field representative for the forthcoming census. You are given a pile of census forms and sent to a postal district of the capital city. You are to knock on doors and interview the most senior person in the home, having him or her fill out a five-page questionnaire that asks for the numbers of people in the household, their ages, sex, and occupation. There are also questions involving conditions of health, arrangements of sleeping quarters, types of kitchen appliances, number of windows, and other details that someone in the government thought it worthwhile to tabulate. You are told that there will be many families where the respondent is illiterate, and you will have to aid her or him in filling out the questionnaire. The district you have been assigned is the most violent in the city, with two to three murders every week and menacing looking young men lounging around the alleyways.

What would you do? Would you knock on doors and try to get your forms filled out? Or, would you sit in your car and fake your visits by filling out questionnaires in what you think is a random fashion? When field representatives do the latter, it is called "curbstoning," invoking the vision of the field representative sitting at the curb, filling out forms, and never going up to the house. Such faking of data has always plagued surveys and censuses. The U.S. Census Bureau and the Bureau of Labor Statistics have developed several ways to detect curbstoning. They reinterview a random subset of the subjects supposedly interviewed. If they suspect a field representative, the random subset will be enriched with a selection of forms from the suspect representative. The questionnaires are run through a computer program that looks for inconsistencies. (Did the family have no income but buy a refrigerator last month?) An attempt is made to identify outliers. One procedure is to take one number from the forms (e.g., family income), order these numbers from smallest to largest, and look at the gaps between them. The gaps should follow a particular distribution, and, if the largest gap is greater than expected, that questionnaire is flagged. The Bureau of Labor Statistics also explored the use of Benford's law. This is the empirical finding that, for numbers greater than 1000, the leading digits tend to be concentrated among the values, 1, 2, 3, or 4.

As with most fakers of data, curbstoners usually produce questionnaires whose data show very little variability (small variance). Another statistical procedure for identifying faked data is to look at the percentage of times the questionnaires from a specific field representative have the same results, the percentage of coincident values. Finally, the questionnaires from two different field representatives can be compared for comparable variances.

But, there are cultural attitudes that can produce bona fide results that get flagged. These fall under the category of blunders, elements from a different statistical distribution. In some African communities, children are not counted as people until they reach maturity, so checks on age distributions will flag those questionnaires. In some Asian countries, community harmony is prized above all else, and respondents get together to provide the same answers. When this happens, the questionnaires that result from such cooperation would have been flagged by these computer checks.

Between 1982 and 1987, the U.S. Census Bureau examined the data they had on field representatives who had been found guilty of curbstoning. They used logistic regression to find aspects of the representative, which were predictive of curbstoning. The most important characteristic was the length of time the field representative had been working for the Census Bureau. Curbstoners tended to be people who had worked as field representatives for less than one year. See Reference [1] for details of procedures used by the Census Bureau and the Bureau of Labor Statistics.

The Pew Research Center in Washington, DC, sponsors surveys that tabulate the opinions and activities of people throughout the world. In January 2016, Noble Kuriakose of Princeton University and Michael Robbins of the University of Michigan published a paper that examined the data from 1100 surveys in Africa (many of them sponsored by Pew), and they concluded that as many as 20% of the studies they looked at were contaminated by large amounts of curbstoning.

The statistical test they applied was developed with a Monte Carlo study. In this study, they set up a "questionnaire" with 100 items. Each item could be responded to by either "yes" or "no." In each run of the study, they sampled 1000 "respondents" by randomly assigning "yes" or "no" with equal probability. In each "survey," they determined the percentage of responses that agreed with another response on two or more of the items. They did this 100,000 times and looked at the percentage of questionnaires that had such duplication. They fit this data to a class of probability distributions widely used in quality control called Weibull distributions.

The Weibull distribution is named after the Swedish engineer, Waloddi Weibull (1887–1979), who rediscovered it in 1951. It had first been discovered by Frechet in 1927, and it disappeared into the sea of mathematical theorems that have been proven but never used. Weibull used it in quality control and published his results in a journal that other engineers read.

In none of the 100,000 runs of the Monte Carlo study did more than 85% of the "questionnaires" have two or more matches. The fitted Weibull distribution predicted that the probability of 85% or more matching is less than 1 in 20. They used this as their test of whether a census was contaminated with curbstoning. That is, if, in a given census, they found that more than 85% of the forms match another form on two or more items, they would declare that census contaminated by curbstoning. Their test flagged 1 in 5 of the 1100 censuses they examined as being untrustworthy. Their paper is Reference [2] of this chapter.

In the previous chapters, we have considered the use of well-established statistical tests for detecting faked data. The Kuriakose and Robbins test does not rest on any well-established psychological phenomenon like digit preference or violations of what should be a uniform distribution, as in the frequencies of least significant digits. It rests on the findings of a Monte Carlo study that worked with an artificial "questionnaire." A real-life questionnaire has more detail than 100 independent "yes" or "no" questions. In many cases, successive questions are related and not independent (as they were in the Monte Carlo study).

The Pew Research Center responded to the Kuriakose and Robbins paper by looking into their data banks of studies where they were fairly sure curbstoning did not occur. They ran a multi-linear regression, looking at the percentage of matching questionnaires as a function of various characteristics of the questionnaire, see Reference [3]. They showed that 85% coincidences on two or more of the questionnaires are a function of the number of items and of their interrelationship. The more questions on the questionnaire, the greater the probability that two forms would have the same answers on two or more questions.

As of this writing, there was no general agreement whether Kuriakose and Robbins had uncovered a real problem. However, studies during the 1970s showed that the official economic and census reports from dictatorships or states ruled by "presidents for life" frequently showed all the characteristics of faked data, digit preference, low variance, and too many equal numbers. This held for economic reports from Soviet Russia and China under Mao Tse-tung, as well as dictatorships like that of Sukarno in Indonesia. Faked official data seems to be particularly prevalent in closed societies. Is this true also of curbstoning?

This example shows one of the weaknesses of statistical models in detecting fraud. You start with a model for error, which you assume is appropriate to the problem at hand. You then generate the probability distribution of some characteristic of the data that should be sensitive to departures from the expected model. Digit preference and lack of variability are well-known marks of fakery, and there are well-established statistical tests to uncover those types of nonrandom patterns associated with fakes.

Kuriakose and Robbins did not use one of the established statistical approaches for detection of fakes. Instead, they proposed a distribution of error that they claimed was appropriate to the problem, and they generated the probability distribution for their test of that property. They failed to show that the property being tested (the percentage of census forms that agreed on two or more items) could, in fact, distinguish between data that was known to have been curbstoned and data that was known to be proper. The Pew researchers believed that the Kuriakose and Robbins model of the error distribution was inappropriate when the census form was very long or when there were related questions in the form.

The human population of a country tends to stop growing exponentially when the nation becomes wealthy enough to spread the benefits of modern medicine on longevity. Families have many children when it is expected that only a few will survive to adulthood and be able to help aging parents. They have fewer children when they know that all or most will survive. Using the fact that wealth is the best contraceptive, demographers have long believed that, as the third world gets more and more wealthy, the total human population will level off at seven billion around the middle of this century.

In spite of the general effect of wealth on population, human population has continued to surge in Africa. Because of this, current estimates are that the world population will level off at 10 billion. Is this true? Has there really been an unexpected surge in African population—or are we seeing the effects of curbstoning?

17.1 SUMMARY

A major problem with censuses and economic surveys is that the field representatives who interview the subjects of the survey may be faking many of the interviews. This is called "curbstoning." The U.S. Census Bureau and the Bureau of Labor Statistics have developed tests that look for outliers and inconsistencies in completed questionnaires, and they contact a random subset of the subjects supposedly interviewed to verify that they were. Kuriakose and Robbins have examined 1100 surveys in Africa, most of them supported by the Pew Foundation. Using their own test developed from a Monte Carlo study, they claim that 20% of the surveys are contaminated by curbstoning. The Pew Foundation has applied multilinear regression to their own surveys and cast doubt on the validity of the test used by Kuriakose and Robbins. There has been an unexpected surge in the population of Africa, causing demographers to increase their estimate of how the world population will grow in this century. Was there a real surge or was it curbstoning?

REFERENCES

1. Swanson, D., Cho, M.J., and Eltinge, J. (2003) Detecting Possibly Fraudulent or Error Prone Survey Data Using Benford's Law, presented at the 2003 Joint Statistical Meetings. Available at: https://www.amstat.org/Sections/Srms/Proceedings/y2003/Files/JSM2003000205.pdf.

2. Kuriakose, N., and Robbins, M. (12 December 2015) Don't get duped: Fraud through duplication in Public Opinion Surveys. *Stat J IAOS*. Available at: http://ssrn.com/abstract=2580502.

3. Simmons, K., Mercer, A., Schwarzer, S., and Kennedy, C. (2016) Evaluating a New Proposal for Detecting Data Falsification in Surveys. Available at: http://www.pewresearch.org/2016/02/23/evaluating-a-new-proposal-for-detecting-data-falsification-in-surveys/.

Glossary

adjusted R^2: Percent of variance accounted for, adjusted to account for the number of elements in the regression.

average: The sum of the observed values divided by the number of observed values. This is the most widely used estimator for the mean of a distribution.

central limit theorem: The conjecture that any errors that can be thought of as the sum of a large number of smaller values can be described as being normally distributed.

consistency: The property of an estimator where the probability of the estimate of a parameter being close to the true value of the parameter increases as the number of observations increases.

dummy variable: An element of a regression, which is 1.0 if some event has occurred and 0.0 if it has not.

false discovery rate: In the "big p little N" problem, the estimated proportion of times "significant" relationships will turn out to be false.

likelihood: The distribution probability formula with the observed values in the formula, so the only unknowns are the symbols for parameters.

log-odds: If p = the probability of some event, then the log-odds are computed as $\log_e(p/(1 - p))$.

maximum likelihood estimators: Estimators of parameters that are derived from maximizing the likelihood associated with the observed values.

mean: The center of the error distribution, which is a parameter that can be estimated from the observed data.

median: A number such that half the observations are less than or equal to it and half are greater than or equal to it. In the theoretical distribution of the error, a median is a number such that the probability of being less than or equal to it is equal to the probability of being greater than or equal to it, 0.50.

minimum variance: A desirable property of an estimator, where the variance of this estimator is less than or equal to the variance of any other estimator.

Monte Carlo studies: Computer studies where data are generated from a specific probability distribution and used to explore the probabilistic behavior of different functions of the data.

normal distribution: A theoretical distribution function for the error that is widely used in statistical analyses because of its properties that lead to relatively easy calculations. If the central limit theorem holds for the problem at hand, the error is normally distributed.

odds: If p = probability of some event, then $p/(1 - p)$ = the odds for that event.

Poisson distribution: A one-parameter distribution that arises from a situation where events are scattered in such a way that the probability of at least one event being captured is a function of the size of observed unit.

R^2: The square of the correlation between observations and the predictions of a proposed model. It measures the proportion of the variance of the observations that is accounted for by the predictions.

randomized response survey: A survey where the percentage of people engaging in an illegal or socially unacceptable behavior is uncovered by having each respondent answer one of two questions at random.

restricted test: A statistical test which has been tailored to be sensitive to a well-defined hypothesis.

robust (robustness): The property of a statistical procedure that allows it to be useful even if some of the mathematical assumptions behind it are not true.

sample standard deviation: An estimate of the standard deviation of the error distribution derived from the observed data.

sample variance: An estimate of the variance of the error distribution derived from the observed data.

standard deviation: The square root of the variance of the error distribution.

stepwise regression: Multilinear regression where there are many candidates for x-variables and a small subset is selected by looking at the percentage of the variance accounted for.

symmetry: A property of the error distribution where the probability of having an error greater than the mean is equal to the probability of finding an error of the same absolute value, but less than the mean.

trimmed average: An estimator of the mean of a possibly contaminated error distribution that orders the observed values and uses only the middle XX%, where XX is determined by what might be expected from a contaminated distribution.

unbiased: The property of an estimator of a parameter such that the mean of the error distribution for that estimator is the value of the parameter.

variance: A measure of the spread of the error distribution. It is the mean of the squared differences between the mean and all other values of the distribution.

Winsorized average: An estimator of the mean of a possibly contaminated distribution that orders the observed values and changes the values that are in the upper and lower XX% to the next value in. The number XX is determined by what might be expected from the contaminated distribution.

Weibull distribution: A distribution which is not symmetric, but its parameters can be easily estimated from the conditions under which the observations have been made. It is widely used in engineering and quality control.

Index